FRONTIER BOTANIST:
William Starling Sullivant's Flowering-Plant Botany of Ohio (1830-1850)

By

Ronald L. Stuckey [1]

and

Marvin L. Roberts [2]

A Contribution from The Ohio State University Herbarium

Department of Botany, College of Biological Sciences

The Ohio State University

Columbus, Ohio 43210

1991

[1] Portions of this paper were read to the Historical Section, Botanical Society of America, Annual AIBS Meeting, Stillwater, Oklahoma, 14 August 1979; the Ohio Academy of Medical History, Annual Meeting, Granville, Ohio, 26 April 1980; and the Historical Section, Botanical Society of America, Annual AIBS Meeting, Columbus, Ohio, 12 August 1987.

[2] Present address: Biology Department, Salem State College, 352 Lafayette College Street, Salem, MA 01970-4589.

Sida, Bot. Misc. No. 6 **ISSN 0883-1475**

SIDA, BOTANICAL MISCELLANY

Founded by Wm. F. Mahler and Barney L. Lipscomb, 1987

Botanical Research Institute of Texas
Herbarium
Southern Methodist University
Dallas, Texas 75275

Sida, Botanical Miscellany, No. 6

DEDICATION

Dedicated to the Memory of

ANDREW DENNY RODGERS III,

who gave us so much of the Sullivant Family History

and to

Professors **EMANUEL D. RUDOLPH** and

JOSEPH EWAN, who have inspired us

in our studies of the history of botany.

Published in

Celebration of the Centennial Meeting
of the
Ohio Academy of Science
Columbus, Ohio, 26-27-28 April 1991

Fig. 1. A pictorial representation of William Starling Sullivant (from Müller, 1875). This illustration is not known ever to have been published in an American work on Sullivant.

Table of Contents

List of Figures

ILLUSTRATIONS NOT REFERENCED

LINE DRAWINGS OF PLANTS

[Editor's Note: Scientific names and foreign words have been italicized in the quotations]

Fig. 2. Monument of Lucas and Sarah (Starling) Sullivant in Green Lawn Cemetery, Columbus, Ohio with medallion of Lucas Sullivant (photograph by Ronald L. Stuckey, November 1981).

Preface

My interest in the history of botany, and especially the botanical accomplishments and associates of William Starling Sullivant, began in 1962, when I was a graduate student in Botany at the University of Michigan, Ann Arbor. At that time, the first book I read in this discipline was Mr. Andrew Denny Rodgers' "'Noble Fellow' William Starling Sullivant." His account of Mr. Sullivant's life impressed me with the manner in which he wove together botanical information, family details, and Franklin County history. Subsequently, I received much encouragement to pursue investigations in the history of botany from Dr. Edward G. Voss, my major professor, and Dr. Rogers McVaugh, curator and senior research professor in the University of Michigan Herbarium. Prof. Joseph Ewan of Tulane University also became a valuable correspondent and supporter of my biohistorical studies.

My first major paper in the history of botany was a detailed account of the botanical work of John Samples, a pioneer botanist who lived most of his botanical life at Urbana, Ohio. That paper was published in the Ohio Journal of Science 66:1-41. 1966. While working on the Samples project, Dr. Clara G. Weishaupt, Curator of The Ohio State University Herbarium, first brought to my attention W.S. Sullivant's 1840 Ohio specimens of flowering plants in the OSU Herbarium. My first letter to Sullivant's biographer, his great-grandson, Mr. Rodgers, author of "Noble Fellow," was mailed in the fall of 1963, and his enthusiastic and informative reply soon followed. It was indeed a great pleasure when Dr. and Mrs. Emanuel D. Rudolph introduced me personally to Mr. Rodgers at their home for an evening dinner in the fall of 1965, shortly after I joined the Department of Botany at The Ohio State University. Both Mr. Rodgers, in his remaining lifetime, and Dr. Rudolph have been a great inspiration to me during my 25 years of research in the History of Botany.

Since 1968, my co-author, Dr. Marvin L. Roberts, who initially worked for me while an undergraduate student, traveled with me to several libraries and herbaria to seek information on Sullivant's work as a phanerogamic botanist. He organized the data from Sullivant's flowering plants in The Ohio State University Herbarium into a catalogue. He has brought to my attention many obscure and noteworthy items, including those published by Bigelow (1843), Crosby (1915), Curtis (1839, 1840), Harshberger (1899), and Wheaton (1882). All of these facets, including writing and editing the manuscript have kept our interest in Mr. Sullivant alive during the past two decades.

Many of my friends have introduced me to memorabilia concerning the Sullivant family and its associated botanical achievements. Dr. Charles C. King of the Ohio Biological Survey, by his slide photographs, first brought to my attention the impressive Sullivant markers in Green Lawn Cemetery. My first visit there was in the spring of 1981, being guided to the site by my good friends, Drs. Milton B. and Mary (Auten) Trautman. Several visits followed with Milton, including one in October 1982, with our mutual friend, Mr. Parker Bauer of Oklawaha, Florida. Other companions during subsequent visits have been Ms. Bette Roth, Ms. Melanie Pratt, Dr. Marvin L. Roberts, and Mr. John W. Frederick.

As part of the program of the Historical Section of the Botanical Society of America, AIBS Meeting August 1987, I led a field trip to the Sullivant family plot in Green Lawn Cemetery, Columbus, Ohio. Over 20 individuals participated in this excursion, and as a keepsake for this event, I prepared an 86-page xerox booklet, *First Botanists of Columbus: A Sullivant Family Scrapbook* (Stuckey, 1987). A follow-up article concerning my research on Columbus' founding family appeared in the January issue of *Columbus Monthly* by Jane Ware (1991).

Ronald L. Stuckey
Professor of Botany
The Ohio State University
20 September 1990

Acknowledgments

Grateful appreciation is extended to the many individuals who have assisted in our studies of William S. Sullivant. Regrettably, not all of these individuals can be named here. We are especially grateful to the late Mr. Andrew Denny Rodgers III, Sullivant's great-grandson and biographer, who shared his vast knowledge with us, and whose personal transcripts of Sullivant's letters to Asa Gray (the originals of which are housed in the Library of the Gray Herbarium, Harvard University) were left to the senior author. Other sources for Sullivant's correspondence are the New York Botanical Garden and the Historical Society of Pennsylvania, Philadelphia (letters to John Torrey); the Missouri Botanical Garden, St. Louis (George Engelmann); the Filson Club, Louisville (Charles W. Short); and the Historical Society of Wisconsin, Madison (Increase A. Lapham). Other libraries consulted are the American Philosophical Society; the Cincinnati Historical Society; the Lloyd Library, Cincinnati; the Ohio Historical Society, Columbus; The Ohio State University, Columbus; the Academy of Natural Sciences of Philadelphia; and the College of Physicians, Philadelphia. Considerable thanks is given to the curators of the herbaria consulted for Sullivant's Ohio vascular plant specimens. These are DWC, GH, MO, NY, OS, and PH. To all of these institutions we are most grateful.

Thanks for assistance in their various capacities go to Mr. J. Perry Edwards, Prof. Joseph Ewan, Mr. John Frederick, Patricia Kimball, Dr. John J. Furlow, Dorothy I. Lansing, M.D., Dr. William Overlease, Mr. Gregory J. Schneider, Dr. Alfred E. Schuyler, and Dr. David H. Stansbery; and to Messrs. Ronald E. Mossman and A. E. Spreitzer, who prepared the photographs for publication. The line illustrations of the plants appearing on pages 47, 48, 52, 55, and 59 are taken from the first edition of "An Illustrated Flora of the Northern United States, Canada, and the British Possessions" by N. L. Britton and A. Brown (Vol. 1, 1896; Vol. 2, 1897; Vol. 3, 1898).

LUCAS SULLIVANT
From a Medallion by J. K. Campbell.

Lucas Sullivant as drawn from a medallion by J. K. Campbell (from his granddaughter, Jane D. Sullivant, 1928).

FRONTIER BOTANIST:

WILLIAM STARLING SULLIVANT'S FLOWERING-PLANT BOTANY OF OHIO

(1830 - 1850)

"A NOBLE FELLOW IS SULLIVANT AND DESERVES ALL YOU SAY OF HIM AND HIS WORKS. THE MORE YOU GET TO KNOW OF HIM THE BETTER YOU WILL LIKE HIM." (Asa Gray, 1857, letter to Sir William J. Hooker, 6 December).

EARLY LIFE--BEFORE BOTANY

William Starling Sullivant (1803-1873) is perhaps best known for his contributions to bryology, the study of mosses and liverworts, which he conducted during the latter half of the nineteenth century. A resident of Franklinton and later of Columbus, Ohio, on the frontier of American life in the mid-1800s, Sullivant (Fig. 1) gained considerable recognition in this country and abroad for his accomplished work on the study of North American bryophytes. His contributions were in the form of exchanges of plant specimens, preparation of sets of exsiccatae, publication of scientific papers and books, and the development of an extensive correspondence with botanists throughout the world.

The definitive biography of Sullivant, prepared by his great-grandson, Andrew Denny Rodgers III (1940), weaves together a detailed account of Sullivant's life, the development and growth of Columbus and the surrounding area, and his botanical contributions, primarily to bryology. Shorter biographies are by Gray (1873, 1874, 1877), Müller (1875), Rodgers (1962, 1969), Smith (1905), Stokes (1914), Wilson and Fiske (1888), and Youmans (1896). As noted by Rodgers (1969), Sullivant and his co-worker, Leo Lesquereux (1806-1889), laid the foundation for the taxonomic study of the moss and liverwort flora of North America in much the same way that John Torrey and Asa Gray did for the study of the phanerogamic flora of North America. Sullivant, under the early influence of Torrey and Gray, really launched his botanical studies by working with the phanerogamic flora of central Ohio, but little information concerning this beginning -- the events, labors, contributions, and accomplishments -- has been discussed and analyzed. From a study of a collection of phanerogamic specimens in the herbarium of The Ohio State University, a restudy of Sullivant's letters to Torrey and to Gray, and a synthesis of information from Sullivant's letters to other contemporary botanists, published sources, and specimens in other herbaria, it is now possible to document further Sullivant's beginning botanical work coincident with his contributions to phanerogamic botany (Stuckey, 1979).

William Starling Sullivant's father Lucas Sullivant (Fig. 2) was a native of Mecklenburg County, Virginia. At an early age he began to master mathematics with the idea of making surveying his profession. At age 22, in 1787, he received an appointment as deputy surveyor from Colonel Anderson, surveyor general of the Virginia Military District. With this obligation Lucas Sullivant and his surveying party journeyed northward through Kentucky into southern Ohio to survey wilderness lands between the Scioto and Miami rivers. During the next 10 years he acquired ownership of large tracts of land along the Scioto River, organized the county of Franklin, and built the first brick house in that town where he lived the remainder of his life (Anonymous, 1901; Jane Sullivant, 1928).

William Starling Sullivant, as summarized from Rodgers (1940, 1969), was born 15 January 1803 to Lucas and Sarah (Starling) Sullivant in the village of Franklinton, now part of Columbus. As a young boy he accompanied his father on some of his short surveying

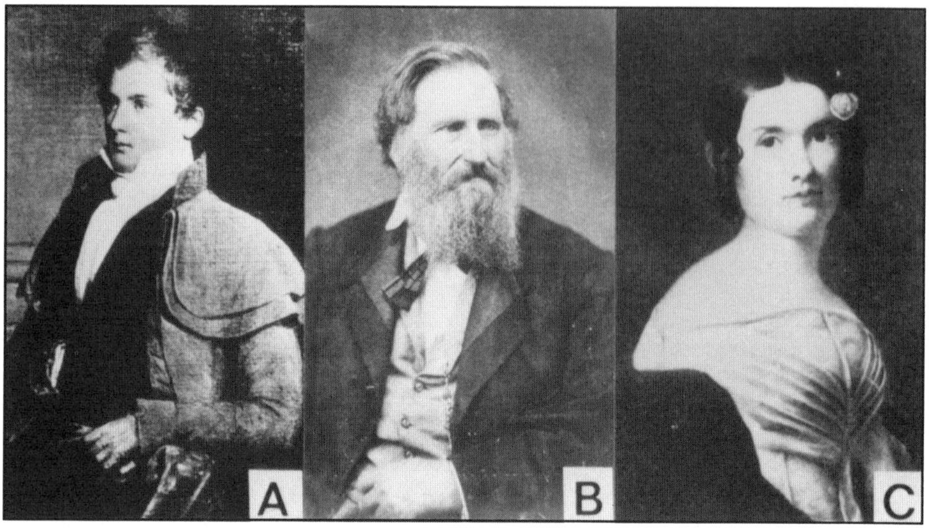

Fig. 3. William S. Sullivant and his two family botanical associates. **A.** William Starling Sullivant (1803-1873) at age 19 or 20. The portrait was painted while Sullivant was a student at Yale College (from Rodgers, 1940, opp. title-page). **B.** Joseph S. Sullivant (1809-1882) (from a photograph in the Photo Archives, The Ohio State University, Columbus; another copy is in the Museum of Zoology, The Ohio State University). **C.** Wedding portrait of Eliza Wheeler Griscom Sullivant (1817-1850) at age 17 (from Rodgers, 1936, p. 97; also appearing in Rodgers, 1940, opp. p. 98).

expeditions. Here the son took his first lessons in woodcraft, surveying, and nature study. At the age of eight or nine years, William was taken to the home of his grandfather to be educated in a celebrated private academy, probably that of the noted educator Samuel Wilson, in Jessamine County, Kentucky. Additional study may have been at Transylvania University in Lexington in 1818, followed by a year completed at Ohio University, Athens, before registering at Yale College in November 1819 and graduating from there in 1823. His courses of study throughout his formal education appear to have been mainly classical, acquiring a proficiency in Latin and Greek, rather than in the scientific disciplines. William planned postgraduate study, but with the death of his father in 1823, he returned to his homestead as a wealthy landowner and businessman. At age 21, he married 16-year-old Jane Marshall on 8 April 1824, but she died the next January following the birth of their daughter. Alone during the decade to follow, William gave foremost attention to the business of managing his large land holdings, mills, stone quarries, banking interests, investments in stage coach, canal, and other companies, while pursuing interests in civil engineering and surveying.

Development of Sullivant's Botanical Interest

The development of Sullivant's interest in the natural sciences and the focus of his attention on the study of botany was believed by Rodgers (1940, pp. 100-101, 105) to have come from the strong influences of his second wife, Eliza Griscom Wheeler, of New York City, and his youngest brother, Joseph (Fig. 3). Rodgers was unaware of the story told by Dr. Asa Horr, medical botanist of Baltimore, Ohio, in Fairfield County, and later botanical companion of William Sullivant. As quoted from Crosby (1915, pp. 163-164), Dr. Horr revealed that:

> One day as Mr. Sullivant and I were alone in a boat on a lily pond, gathering lilies and searching for other water plants, he related to me the incidents that led him to the study of botany. He said: "When a young man, by inheritance, I became the owner of the farm on which my present home is situated. I had no plan of life and was rather inclined to be gay and associate with young men fond of a good time. One day I had four of them at my home for dinner and a little jollification. Looking out of a window that showed the pasture in the landscape, I saw a man walking slowly along, closely watching the ground, [p. 164] occasionally stooping down as if to pick up something, stopping to examine it and then putting it in a tin case which was suspended by a shoulder strap at his side."
>
> "I wondered what the man found of so much interest in the pasture, and said to my company: 'Boys excuse me for a little while! I see a man down in my pasture and I must go down and see what he's doing there.' So I left them and went to the pasture. I found a man somewhat advanced in years who explained that he was studying the flora of the state, and had already found in my pasture some new plants not yet described, that he would add to the list. I staid with him till near dinner time, asked him to take dinner with me and he consented. I wanted to see more of him, and if he were not accustomed to our style of living, it might be some fun for the boys as his clothing was suited to his work. When seated at the table, his dignified bearing and intelligent conversation kept my other guests as attentive listeners, with no thought of making fun at his expense. I asked his permission to accompany him the rest of the day, and adjourned the frolic with my gay young friends. That afternoon opened a new world to me and led me to become a student of nature."

Although the name of the man in the pasture and the date of the event are unknown, it would appear from the context of the story that the event occurred before 1834, the year of his second marriage. It evidently was about the time of this marriage to Eliza Griscom Wheeler that Sullivant's enthusiasm for the study of the sciences--ornithology, conchology, and preeminently botany--surfaced in his life. His early botanical efforts involved the study of the local flora by identifying the flowering plants and making an herbarium of them. Among the plants that Sullivant found of most interest were the grasses and sedges. Because of their tiny inconspicuous flowers, they were a great challenge for him to identify.

Sullivant's youngest brother, Joseph S. Sullivant (1874) also took credit for an arousal of these early botanical aptitudes in William. Joseph had already made some progress in the sciences, for he had found the richest source for plants in the immediate vicinity of the white brick mansion, "Sullivant Hill" (Fig. 4), where William's family lived in the country about three miles west of Columbus. Some of Joseph S. Sullivant's natural history study resulted in a *Catalogue of Shells, Fossils, Minerals, and Zoophytes* (1838). In a review of this paper, Alva Curtis (1839), founder of the Literary and Botanical-Medical Institute of Ohio in Columbus, called attention to his extensive collections of these organisms as well as "several hundred specimens of dried plants, collected principally in this country . . . ."

> W. S. Sullivant, Esq., is equally devoted to the study and cultivation of natural history. We have not examined his plants; but a catalogue [apparently an unpublished forerunner of Sullivant's 1840 Catalogue] he furnished as a guide to exchange, in our eastern tour, together with the exchanges and purchases there made for him, will swell his herbarium [Fig. 5] to size and interest not equalled in the west, or scarcely surpassed in the Union. We only add, that the politeness and liberality of these gentlemen, to their scientific or curious visitors, are equal to

Fig. 4. "Sullivant Hill," home of William S. Sullivant, three miles west of Columbus, at 1960 West Broad Street. The property was sold to the State of Ohio for the construction of a state psychiatric hospital (from Rodgers, 1940, opp. p. 100).

Fig. 5. Location of William S. Sullivant's herbarium at "Sullivant Hill" (from Rodgers, 1940, opp. p. 100).

> their immense power to gratify research and curiosity; and that the stranger who does not visit these interesting cabinets, and catch somewhat of the scientific enthusiasm of their worthy proprietors, has missed no small portion of the interest of a visit to Columbus.

These plant specimens would have been collected before or during the 1838 season (Fig. 6). The earliest dates seen on any of Sullivant's plants is 1837 for some specimens of *Carex* in the herbarium of the Missouri Botanical Garden.

Among local individuals who assisted Sullivant in his early work on the flowering plants of central Ohio, as mentioned above, were his brother Joseph S. Sullivant and his second wife Eliza Griscom Wheeler Sullivant. Prominent among other local citizens in Sullivant's circle of botanical friends by 1840 were four young practicing physicians: Dr. Asa Horr of Baltimore in Fairfield County; Dr. John Milton Bigelow of Lancaster in Fairfield County; Dr. Jonathan Roberts Paddock of Worthington; and Dr. Ichabod Gibson Jones of Columbus, both in Franklin County. These close medical associates made collections of plants and contributed to the knowledge of botany of central Ohio. In so doing, they may have gained more botanically from Sullivant than he did from them (Stuckey, 1987a, 1988; Appendix I). All of this activity was of local concern and interest, but William Sullivant sought outside help and recognition for his work.

EARLY INFLUENCE OF TORREY AND GRAY

How Sullivant came into correspondence with Drs. Torrey and Gray, then the botanical leaders in North America, is largely based on conjecture by Rodgers (1940, p. 104), who believed that Sullivant met Dr. Torrey while on business in New York, and that through Dr. Torrey he met Dr. Gray. Rodgers reproduced what he considered to be the first letter to Torrey, dated 18 December 1838, asking for a subscription to the *Annals of the Lyceum of Natural History* and the purchase of books and two microscopes. The tone and contents of this letter, however, would hardly be expected in an opening letter to a new botanical correspondent or perhaps even to one already of some acquaintance. A recently discovered letter to Torrey, dated 26 November 1838, proves actually to be the first letter Sullivant wrote to Torrey. It reveals that Sullivant originally opened the correspondence with Gray seeking dried specimens of grasses and sedges, and that Gray urged Sullivant to write to Torrey regarding the matter. A large portion of Sullivant's first letter to Torrey follows:

> Columbus Ohio Nov 26th 1838
>
> Doctr John Torrey --
>
> Dear Sir -- I was much gratified with receiving a letter from Doctr Gray just previous to his departure for Europe. From reference made to you in his letter I take the liberty of addressing you. Dr. Gray wrote to me in consequence of an application I made to him for his Setts of Gramineae & Cyperaceae of which he had none on hand complete. He very kindly mentioned that it had been his intention to send me specimens of those orders, but that his time having been so much preoccupied he could not do it until his return. At the same time he expressed a hope that you would find leisure to prepare a parcel for me. I fear this is making an unsurmountable request of you -- but my only apology is a strong desire to study those orders which I find exceedingly difficult and unsatisfactory in the absence of proper Books and facility of consulting Herbaria. Authentic specimens of those orders would be highly prised by myself and several of my friends who are now zealously attending to the Botany of this section of the country. Of the above orders I have collected only about 100 species and named them according to the best lights I could get--of some of them I

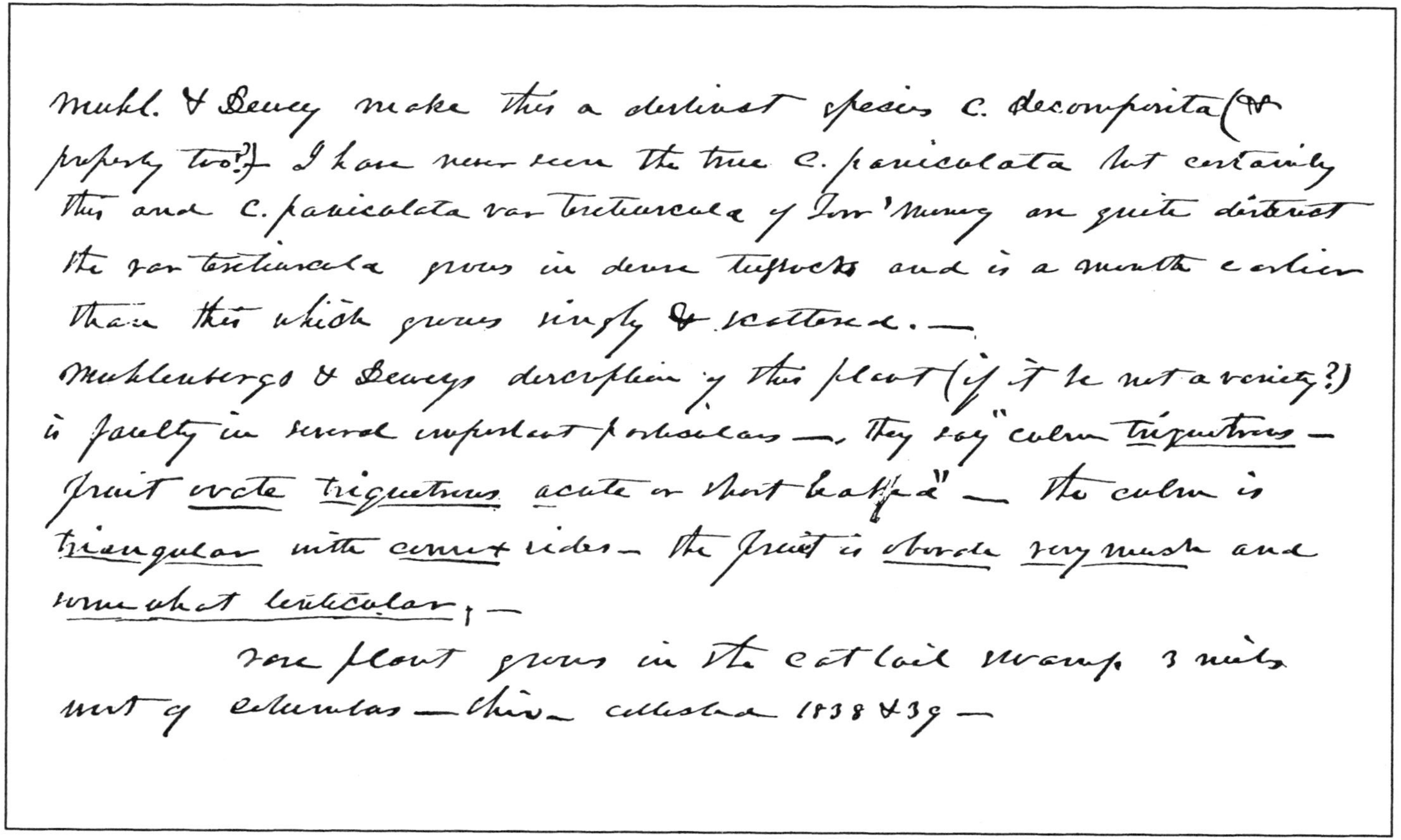

Muhl. & Dewey make this a distinct species C. decomposita (& properly too?) I have never seen the true C. paniculata but certainly this and C. paniculata var teretiuscula of Torr' Monog are quite distinct the var teretiuscula grows in dense tussocks and is a month earlier than this which grows singly & scattered.—

Muhlenbergs & Deweys description of this plant (if it be not a variety?) is faulty in several important particulars—they say "culm triquetrous—fruit ovate triquetrous acute or short beaked"— the culm is triangular with convex sides— the fruit is obovate very much and somewhat lenticular;—

Rare plant grows in the cat tail swamp 3 miles west of Columbus—this collected 1838 & 39—

Fig. 6. William S. Sullivant's handwriting on an extensive label of his specimen of *Carex decomposita* (MO).

> feel doubtful--nearly all the other flowering plants in this vicinity I have collected--and as Doctr Gray mentions that specimens of the more rare and difficult plants would be acceptable I will with pleasure forward to you any thing of the kind we come across--I shall send to you shortly some 10 or 15 of that kind and should you want any other plants we have in this vicinity to replenish your suits I would be glad to send them to you . . . .

After discussing his intention to study mosses, Sullivant asked that Torrey arrange for Gray while in Europe to purchase for him books on the mosses and liverworts, as well as a "very superior microscope." He concluded with a request to exchange botanical specimens, saying:

> I wish to increase my Herbarium as much as possible -- Should you know of any Gentleman disposed to make exchanges with the plants of their section for those of this state, I should be glad to go into the arrangement -- I hope soon (with what I have now on hand) to have a pretty good stock for that purpose -- it might suit some to have the land and water shells of this vicinity for their plants -- if so I have good collections of them to exchange for plants -- please excuse this troublesome letter and in any thing I can be of service to you command me freely --
>
> Yours very Respectfully
> Wm. S. Sullivant

Sullivant's enthusiasm for the study of the grasses and sedges prevailed throughout 1839. On 10 January Sullivant wrote Torrey that the specimens of Gramineae and Cyperaceae sent to him by Torrey gave much pleasure, and on 25 April, he sent a small parcel of flowering plants to Torrey to be named and wrote that more collections would be made during the summer season. "To Cyperaceae and Gramineae I shall give special attention -- they are favorite orders with me." Torrey soon replied with comments, which undoubtedly gave encouragement to Sullivant, and in his 9 June letter, Sullivant said: "Your remarks on the phan[er]ogamous plants I sent you were very acceptable." He also sought more specimens of grasses and sedges from Torrey, noting in his letter of 23 September that "These are difficult to be procured many who attend to Botany do not collect them." On 12 December, Sullivant acknowledged the safe arrival of a truly rare and valuable collection of plants from a part of the country unknown to him. In this package were grasses and sedges of exceedingly great interest, including some 50 plants he had not seen before. To facilitate the study of the genus *Carex*, Sullivant (undated) prepared an Index to the volume and page numbers for the descriptions of the species published by Chester Dewey (1824-1835) in Benjamin Silliman's periodical, the *American Journal of Science and Arts*. The handwritten, 2-page *Index* was mounted in Sullivant's copy of John Torrey's *Monograph of North American Cyperaceae* (1836), presently catalogued and kept in the main library of The Ohio State University, Columbus.

In his correspondence of 1839 with Torrey, Sullivant mentioned making frequent exchanges of plants with his local botanical friends; how imperfectly known some of the plants were, particularly members of the Compositae; certain rare species that he found, including *Erysimum arkansanum, Aronia alnifolia*, a *Melica*, and a *Lonicera*; and sending a box of 500 plants, the latter containing specimens chiefly of the genera *Aster* and *Solidago*, which he had not been able to determine to his satisfaction. In the letter of 12 December, he urged Torrey and Gray, in their spare time

> . . . to look over all the [plants sent earlier] . . . & make the necessary corrections[,] it would be quite a favour to me. -- As I mentioned [earlier] . . . it is my intention (as soon as I hear about them) to have printed a list of all the plants around here (near 1000) that I have collected myself -- Although a small [item] it may not be an unacceptable service to our botanical friends.

> -- My collections here will afford a pretty fair representation of the Flora of the central parts of this state -- a number of them are not catalogued in Riddells synopsis [(Riddell, 1834-1835)].

By this time Sullivant's *Catalogue* (1840) of the vascular plants of central Ohio was well in the making, and in his 24 February 1840 letter to Torrey he informed him that ". . . a week or two since I sent you & Dr Gray copies of my *Catalogue of plants* . . . ."

SULLIVANT'S CATALOGUE (1840)

Sullivant's *Catalogue* (1840), his first book, is a taxonomic list of the vascular plants of Franklin County and nearby locations in central Ohio (Fig. 7). Its primary objective, rather than being a definitive local flora, was to assist him in obtaining exchanges of plant specimens with other botanists, a procedure that is invited in the preface. Over 770 species, however, were listed; the scientific nomenclature followed volume one of Torrey and Gray's *Flora of North America* (1838). Common names were also listed and 46 species were marked as non-indigenous, although some additional ones should have been given that designation (Appendix II). Notes and comments on abundance, habitats, localities, references to literature, and/or mention of individuals who observed the plants in the state were noted for 41 entries. Among local individuals acknowledged for assistance were Joseph S. Sullivant, Lyne Starling, Jr., Colonel A. J. McDowell, John Rowe, T. Roberts, John Samples (cf. Stuckey, 1966), and Thomas G. Lea (cf. Braun, 1934).

In the custom of the time, contemporary reviewers of Sullivant's *Catalogue* (1840) gave more praise to the author as an individual and his herbarium, rather than to the publication itself. Curtis (1840) was impressed with

> . . . the urbanity and liberality with which they [the Sullivant brothers] contribute to the amusement and instruction of those who visit their cabinets or deal with them in exchange. We presume there is not a college west of the mountains supplied with . . . any thing like so valuable an herbarium as that of W. S. Sullivant . . . Besides the catalogue here mentioned, he has a most extensive and beautiful collection of plants from other parts of America, and from Europe, Asia and Africa, which are well worthy of the attention of the scientific as well as the curious.

Dr. Daniel Drake (1840), the eminent physician of the midwest and promoter of the natural sciences in Cincinnati, referred to Sullivant as

> . . . a gentleman of leisure, but not of laziness; and although not under the necessity of being active in business, is active--even laborious, in collecting and studying the plants of his neighborhood. Would that more of our sons possessed his taste.
>
> . . . .
>
> We trust that Mr. S. will continue his labours, gradually extending his field of observation, and that many other floras will at no distant time, be the consequence of his example. It is only by their aid, that a general and complete catalogue of our plants can be made out. Such a work is much wanted , ...

According to Rodgers (1940, p. 118), Sullivant's first book

> . . . was not a difficult one to print, not requiring especial technical skill, and so a Columbus printer of that time, Charles Scott, was employed . . . The

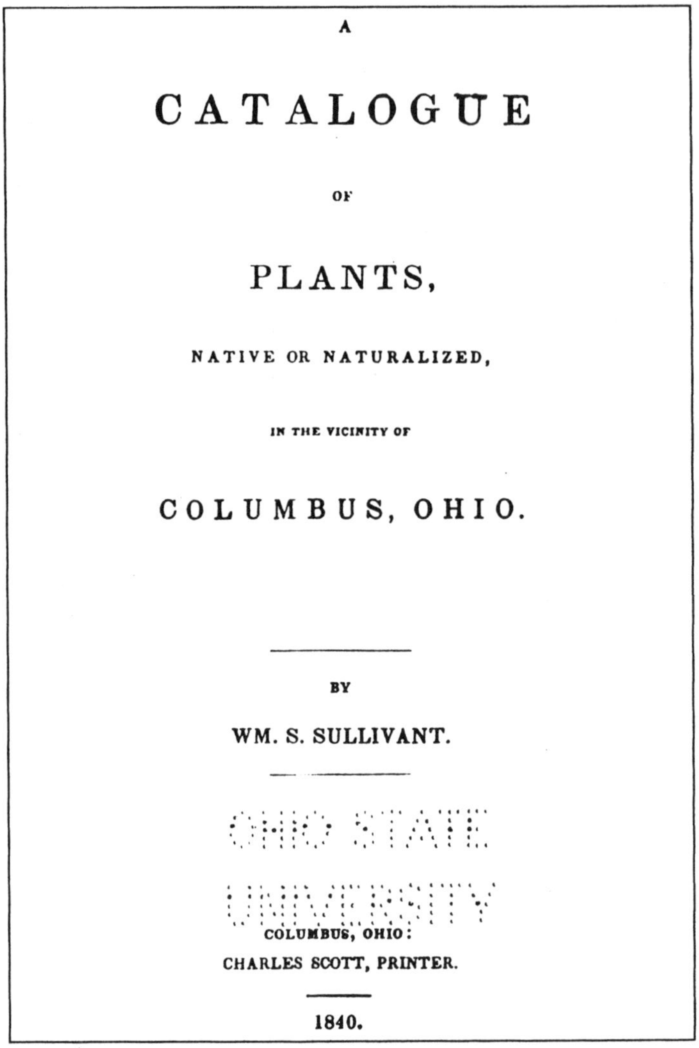

A

CATALOGUE

OF

PLANTS,

NATIVE OR NATURALIZED,

IN THE VICINITY OF

COLUMBUS, OHIO.

BY

WM. S. SULLIVANT.

COLUMBUS, OHIO:

CHARLES SCOTT, PRINTER.

1840.

Fig. 7. Title page of Wm. S. Sullivant's *CATALOGUE OF PLANTS ... IN THE VICINITY OF COLUMBUS, OHIO.*

catalogue of central Ohio plants, however, brought Sullivant some notice but by no means the recognition he sought.

Publication of the *Catalogue* allowed Sullivant to expand his botanical correspondence, as for example, the offer and follow-up of an exchange of plants by Increase A. Lapham of Milwaukee, Wisconsin, as revealed in letters of 31 March, 25 April, and 12 October 1840 (Lapham, 1811-1875).

As the flora of central Ohio was restudied in the years following Sullivant's death, his *Catalogue* took on additional importance. For example, Selby (1899) published a list of flowering plant families of Franklin County and recorded the numbers of native and non-indigenous species in each family. These data were then compared to similar data compiled from Sullivant's *Catalogue* (1840). With respect to the total number of species, Sullivant had 771 and Selby recorded 1142, for an increase of 371 species, or 32%. It was also determined that 68 of the 771 species that Sullivant listed were non-indigenous, representing 8.8% of the flora; whereas 60 years later the percentage of the non-indigenous flora had nearly doubled to 16.1%, based on 184 of 1142 species. A study seven years earlier by Lazenby (1892), comparing Sullivant's *Catalogue* (1840) with a list of Franklin County plants by Selby and Craig (1890), showed that the foreign floristic element had increased to 15%. These independently published studies showed similar increases in the non-indigenous element.

OBSERVATIONS ON VARIATIONS AND HYBRIDIZATION AMONG PLANTS

Sullivant's keen observations of plants in the field are well documented in his letters to Torrey and to Gray. He noted intermediate plants between species in the genera *Carex, Helianthus, Silphium,* and *Thaspium.* Concerning a particular *Carex*, he wrote to Gray on 20 April 1841:

> You ask "can it be a hybrid between *C. pubescens* & *C. gracillima*" -- the same thing has frequently occured to me. -- But I still think it is a distinct species. I found it in many different places, within the space of 10 acres, upon wet land (cleared of heavy timber within the last 5 or 6 years and never cultivated), growing among *C. pubescens*[,] *C. gracillima* & a dozen other of our most common carices, which are always to be met with in such localities. -- I am not aware that carices ever do hybridize -- if such were their tendency I should suppose there would be frequent cases of it -- they grow in such immense quantities together, intermixed among one an other. -- owing to the close resemblance of this *C. pubes.* & *C. gracil.* it might be easily overlooked. -- plants of this kind have been but little attended to in the central parts of this state, as far as I know, except what little I have done.

Late, in July of that year in an undated letter, he again wrote to Torrey commenting on the *Carex*, saying:

> The *Carex* supposed to be new I am afraid will turn out a hybrid between *C. gracillima* & *C. pubescens* -- It appeared this season in the same place in increased quantities -- but of many hundreds carefully examined not a single nut came to maturity -- This is a very suspicious circumstance -- But is it conclusive on the subject? -- did ever a case of the kind fall under your notice? --

And during the next month, on 18 August 1841, he told Gray that he would send Francis Boott of London, England, all of his Carices. These specimens would be of value to Boott, since Boott, who was to become a world's authority on this difficult genus of sedges, was then preparing a monograph. Concerning the possible hybrid, Sullivant further added:

> Whether the *Carex* is or is not, a hybrid I am unable to decide -- it is perfect & uniform in all its characters, except that, in the only locality I have yet found, the fruit is abortive. -- this circumstance was very suspicious in my mind -- But Dr Torrey tells me it is not conclusive that the same occurs in others e. g. *Stipata* -- & that he thinks it a new species -- considering all things it is perhaps safer to view the abortion in this case as accidental & not owing to Hybridity (a troublesome fellow to be among carices) which as far as I can learn has not been detected in this genus.

During the following year, Boott (1842) named and described the plant as *Carex sullivantii.* Its Latin description was published as a footnote in Asa Gray's (1842) *Notes of a botanical excursion to the mountains of North Carolina,* . . . . The plant was illustrated with a line drawing (Fig. 8) by Dewey (1845). Twenty years later, when Boott (1862) published the third part of his monograph of the genus *Carex*, he maintained *C. sullivantii* as a distinct taxon, noting that:

> Mr. Sullivant found it growing with *C. pubescens* and *C. gracillima*, and thinks it may prove to be a hybrid between them. In a note to me he says: "It has never been found by any one, that I am aware of, but myself, and only in one small spot. I transplanted it into my garden, watched it for several years, but never could find mature achenia."

Subsequent writers, Fernald (1950), Howe (1881), and Mackenzie (1935), have considered the taxon to be of hybrid origin, between *C. pubescens* (now *C. hirtifolia*) and *C. gracillima*, just as Sullivant had considered it. Based on today's knowledge of field conditions under which hybrid plants are formed, the newly man-disturbed open habitat in the once forested region of central Ohio, as described by Sullivant, would certainly have been conducive to the formation of hybrid plants. Sullivant's source of knowledge and ideas about hybridization are not known, but they may have been learned through reading about the subject in books. His comments, however, predate by 25 years the classic work on hybridization by Gregor Mendel, and by 35 years the work on the effects of cross- and self-hybridization in plants by Charles Darwin.

Considering some observations in the genus *Helianthus*, Sullivant wrote in an undated letter in early 1842 to Gray that a particular *Helianthus*

> . . . struck me as something not common, it must be rare -- I never met with [it] but in one spot in the Darby Plains.

Fig. 8. Illustration of CAREX SULLIVANTII Boott (from Dewey, 1845).

Torrey and Gray (1842, p. 324) named and described it as *H. cinereus* ß ? *sullivantii*. The plant was later reported as a hybrid between *H. mollis* and *H. occidentalis* subsp. *occidentalis*, referred to as *H. X cinereus* (Jackson and Guard, 1957; Heiser, 1969). Another species of Sullivant's concern was his new milkweed, *Asclepias sullivantii*, which was named by Dr. George Engelmann (1848) of St. Louis. Sullivant wrote to Torrey on 25 July 1847: "I always thought it new but could never get you & Gray to think so." Sullivant earlier had called Gray's attention to it as a new species in letters of 25 November 1840 and 15 October 1846, but neither of them ever recognized it as a new taxon.

DISCOVERY AND PUBLICATION OF PLANTS NEW TO SCIENCE

Sullivant, writing to Gray on 25 November 1840, was indeed much pleased that some of his flowering plants were new. As revealed in his letters to Gray in 1840 (9 November) and 1841 (8 January and 6 February), and as explained in considerable detail by Rodgers (1940, pp. 125-130), Sullivant identified himself as a "neophyte" and expressed a lack of confidence in his abilities to describe his newly discovered species. He urged that Torrey and Gray name the plants and publish the descriptions. But they insisted that he prepare the publication, and after encountering some initial difficulties, Sullivant received their continual help; his paper appeared in *Silliman's Journal* in April of the next year (Sullivant, 1842). In the short article, Sullivant described three species as new to science, *Arabis patens, Fedia umbilicata,* and *Eleocharis compressa*, and included comments on other rare species, *Valeriana ciliata, Erysimum arkansanum*, and *Eulophus americanus*. Subsequently, another species of *Fedia* was discovered and described as *F. patellaria* in Gray's first edition of his *Manual* (Sullivant, 1848). The most recent study of the latter species places it as a form of *Valerianella umbilicata* (Eggers-Ware, 1983).

Because Sullivant had established an extensive exchange of plants, other botanists critically studied his specimens and found new taxa among them that were subsequently named and described. In their *Flora of North America*, Torrey and Gray (1838-1843) credited Sullivant with specimens representing 15 taxa (Appendix II), of which names and descriptions were published for three new species and four new varieties. Later, they recognized one of the new species, *Saxifraga sullivantii* (Torrey and Gray, 1840, p. 575), as belonging to a new genus, which they named in his honor as *Sullivantia*, and provided the epithet *ohionis* to create the species name *Sullivantia ohionis* Torrey and Gray, as discussed by Gray (1842). Engelmann (1848, 1866) described a milkweed, *Asclepias sullivantii*, and two rushes, *Juncus brachycarpus* and *J. canadensis* var. ß *brachycephalus*. Boynton and Beadle (1901) named a black-eyed susan, *Rudbeckia sullivantii*. Many of these newly described species are inhabitants of wet prairies and calcareous cliffs, both habitat types that previously had not been well explored in eastern North America. Some of these species are western, occurring at the edge of their range in central Ohio.

Sullivant's early interest in the grasses and sedges resulted in a large number of collections of these plants. Specimens of *Carex* were of exceeding value to Francis Boott, who, as noted above, received many specimens from Sullivant and named one, *Carex sullivantii*, in his honor (Boott, 1842). In his scholarly publication on *Carex*, Boott (1862) named five taxa from Sullivant's specimens as new to science. An annotated list of 22 phanerogams originally described from Sullivant's plants (Appendix III), in addition to comments on six rare species that he discovered (Appendix IV), totals 28 taxa of special significance among Sullivant's collections of flowering plants.

HISTORY OF NAMING OF SULLIVANTIA

Sullivant's knowledge of his namesake, *Sullivantia sullivantii*, has been preserved primarily in letters to Torrey and to Gray at the Gray Herbarium rather than as extensive notes on herbarium labels. It appears that Sullivant first discovered the species in 1839, according to a letter of 6 January 1840 to Gray:

> I much regret I am unable to furnish you the fruit of the new saxifragaceous plant it was barely in flower when I found it last summer -- in Highland County in this state 80 miles south of this place -- the only place I have met with it -- growing in the crevices of very high limestone cliffs over Paint Creek. -- for want of a better name I supposed it at the time to be the *Tiarella bracteata* as given in Eaton intending to give it a better examination at home -- It appears to belong to *Heuchera* except that the stamens are not exsert as is usual in that genus. -- in flower about the last of May -- I shall attend to procuring fruit for you next season. -- In this letter you will find some of the best flowers I have of it.

On 26 April 1840, Sullivant followed with a letter to Torrey stating:

> The fruit of the saxifragaceous plant will be ready for collecting between 1st & 10th of June -- but I hope the specimens of immature fruit I sent you in my letter will enable you to determine the genus --

In answer to a letter from Gray, who was then suggesting a name for the new species, Sullivant wrote on 31 July 1840:

> The specific name of the *Saxifraga* is quite gratifying, -- I have laid in a large supply of that plant in all of its stages, the fruit bears you out entirely in placing it among the saxifraga. -- in the first parcel I will send you plenty of it.

By this time, however, based on Sullivant's immature flowering specimens sent earlier, Torrey and Gray (1840, p. 575) had already described the plants as a new species honoring Sullivant with the epithet *sullivantii* and placing the plant in the genus *Saxifraga*, but with question. After Sullivant had sent some fruiting material and after Gray had replied, Sullivant, in his return letter of 6 February 1841 to Gray, inquired:

> You say that there is something in the fruit of *Sax? Sullivantii* "that looks a little curious" -- I have examined the saxifrages but little, & cannot see that the fruit in the present case has any thing peculiar except the ascending seed & that may not be unusual for aught I know. --

In the meantime, after studying the specimens with mature fruit sent by Sullivant, Torrey and Gray decided that the *Saxifraga sullivantii* belonged to a distinct, yet undescribed, genus. In a follow-up letter of 20 April 1841 to Gray, Sullivant said:

> I have no doubt *Sax? Sulliv.* can afford as <u>good</u> generic characters as are to be found in most of the related genera. -- and on the subject of making it a *Sullivantia* -- <u>of course I cannot refuse</u>! i.e. if you think it will <u>stick</u>.

Gray replied in early September 1841, and Sullivant returned a letter of 9 November acknowledging many pleasant items received from Gray, including "nothing more so, than that you are quite satisfied with *Sullivantia Ohioënsis*" as the name for the new saxifragous plant from Ohio. Gray (1842, pp. 21-22) made the name official in a note in *Silliman's Journal*:

> . . . the authors of the Flora [of North America], having received fruiting specimens of this interesting plant, do not hesitate to remove it from the genus [*Saxifraga*] to which it was provisionally appended, and to dedicate it

> to their esteemed correspondent, the promising botanist [W. S. Sullivant] who discovered it.

Following this statement is the formal designation of *Sullivantia* Torrey and Gray as the new genus.

Sullivant certainly was pleased that the saxifragous plant was a new genus and that Torrey and Gray honored him with the generic name *Sullivantia* and the new specific epithet *ohionis*, the state where he lived. Sullivant called the plant his namesake, and he sought to learn more about it by attempting to cultivate it in his garden. He wrote to Gray on 1 July 1842:

> . . . I have just returned from the locality of that plant & brought with me 2 or 3 doz. pots of living plants & 100 or so specimens! I shall send you a good share of them to scatter - some of the pots I intend for the Cambridge garden --

By the end of the year, Sullivant was not having much success cultivating the *Sullivantia*. In a letter of 9 December 1842, he told Gray:

> My name-sake *(S. O.)* (<u>and</u> <u>I</u> <u>am</u> <u>sorry</u> <u>to</u> <u>say</u> <u>it</u>!) shows rather in an unaccommodating disposition -- and not much discrimination withal -- as it evidently prefers the barren rocks of Highland [County] to our fertile Scioto vallies -- twice the removal has been unsuccessful, owing, may be, to the time, Midsummer -- It shall have an other trial next spring with all possible precautions -- come it must, however, *nolens volens*! --

The next spring, in a letter of 19 April 1843 to Gray, Sullivant described the situation:

> Four small pots of *Sullivantia Ohionis* -- all that are alive out of several doz. brought home last summer -- These look well -- I am inclined to attribute the loss of so many of them to the unsuitable season (July) at which they were removed -- The finest specimens of this plant I observed grew in the debris of limestone rocks at the base of the perpendicular cliffs forming the sides of <u>shady</u> <u>moist</u> ravenes -- Evry thing left here in fine condition & if the box reaches you as soon as it should, all I think will live.

During 1843, Sullivant took a different approach and made living collections in the fall of the year: "I sent a man down to Highland Co. to day after the *Sullivantia* --, " he wrote in a letter to Gray of 12 October. By 7 December 1843, in another letter to Gray, he noted: "The Box of Sullivantias &c you have received --" and then gave some advice learned from experience while working with the plants: "the Sullivantias must have plenty of shade and moisture to succeed well --" From both the dried herbarium specimens and the living plants that Sullivant had sent to Cambridge, Gray was able to have prepared full-length, detailed descriptions, wood engravings, textbook drawings, and a fine full-page illustration drawn by his artist, Isaac Sprague. The illustration (Fig. 9) was published in the *Chloris Boreali-Americana* (Gray, 1848). The distribution of *S. sullivantii* in Ohio has been mapped (Fig. 10).

A sometimes overlooked item in the history of the naming of *Sullivantia* is contained in the biography of Sullivant by Rodgers (1940, pp. 130-134). As documented from Sullivant's own specimen labels and letters to Torrey and to Gray, it is clear that the genus was named for Mr. Sullivant. No evidence has been forthcoming to support the idea that the second Mrs. Sullivant, formerly Eliza Griscom Wheeler, was the original discoverer of the *Sullivantia* (King, 1981). Nowhere in the materials examined during this study did Sullivant give her any specific credit for this discovery, even though after her death, Sullivant gave her some general credit when he wrote Asa Gray stating that ". . . she was my constant companion & coworker for the last fourteen years . . ." (letter of 3 November 1850, not sent until 12 October 1851; published in Rodgers, 1940, p. 203).

To accommodate the *Code of Botanical Nomenclature*, not in existence in Sullivant's early years of botanical work, Britton (1913) restored the original epithet, *sullivantii*. Accordingly, the correct name of this species is *Sullivantia sullivantii* (Torrey and Gray) Britton. The most recent taxonomic study of the genus has been received during publication: Douglas E. Soltis. 1991. A revision of *Sullivantia* (Saxifragaceae). Brittonia 43:27-53.

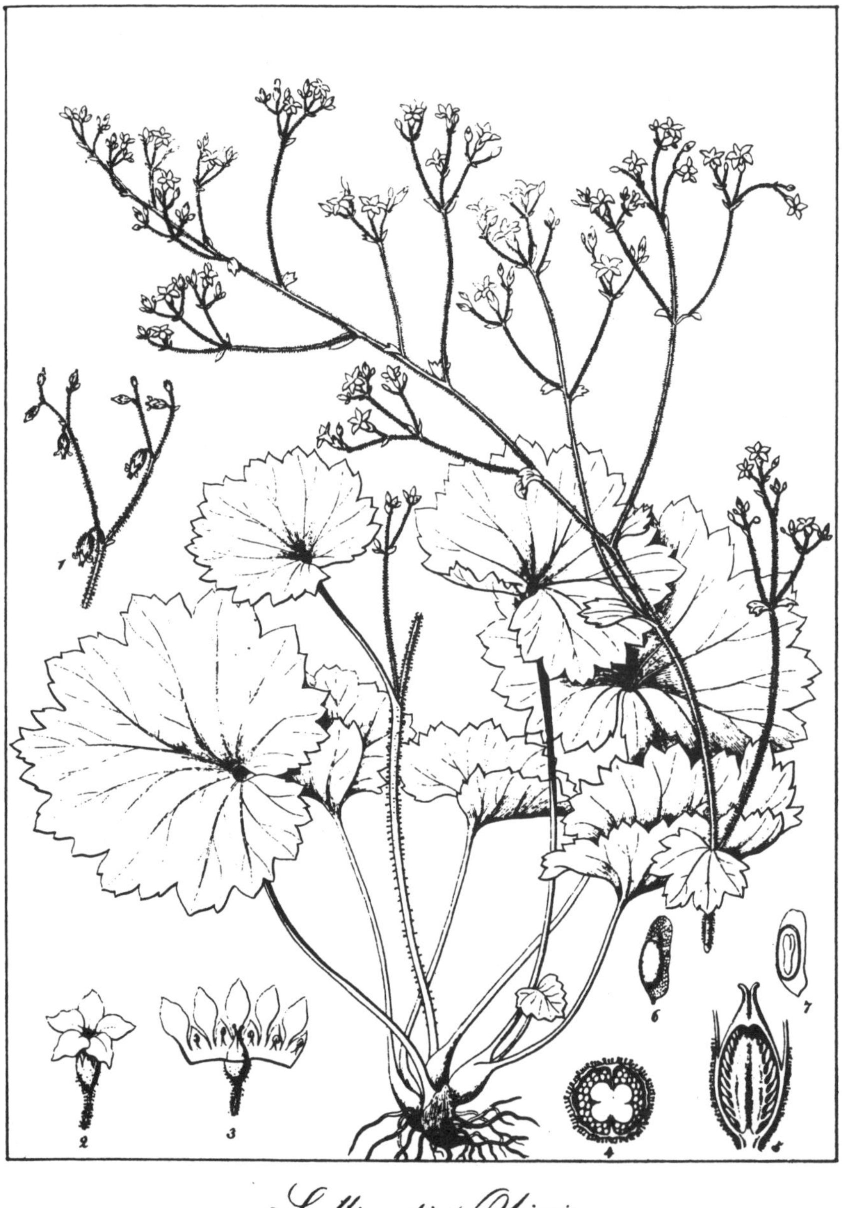

Sullivantia Ohionis.

Fig. 9. Illustration of *SULLIVANTIA SULLIVANTII* (Torrey and Gray) Britton, as drawn by Isaac Sprague (from Gray, 1848).

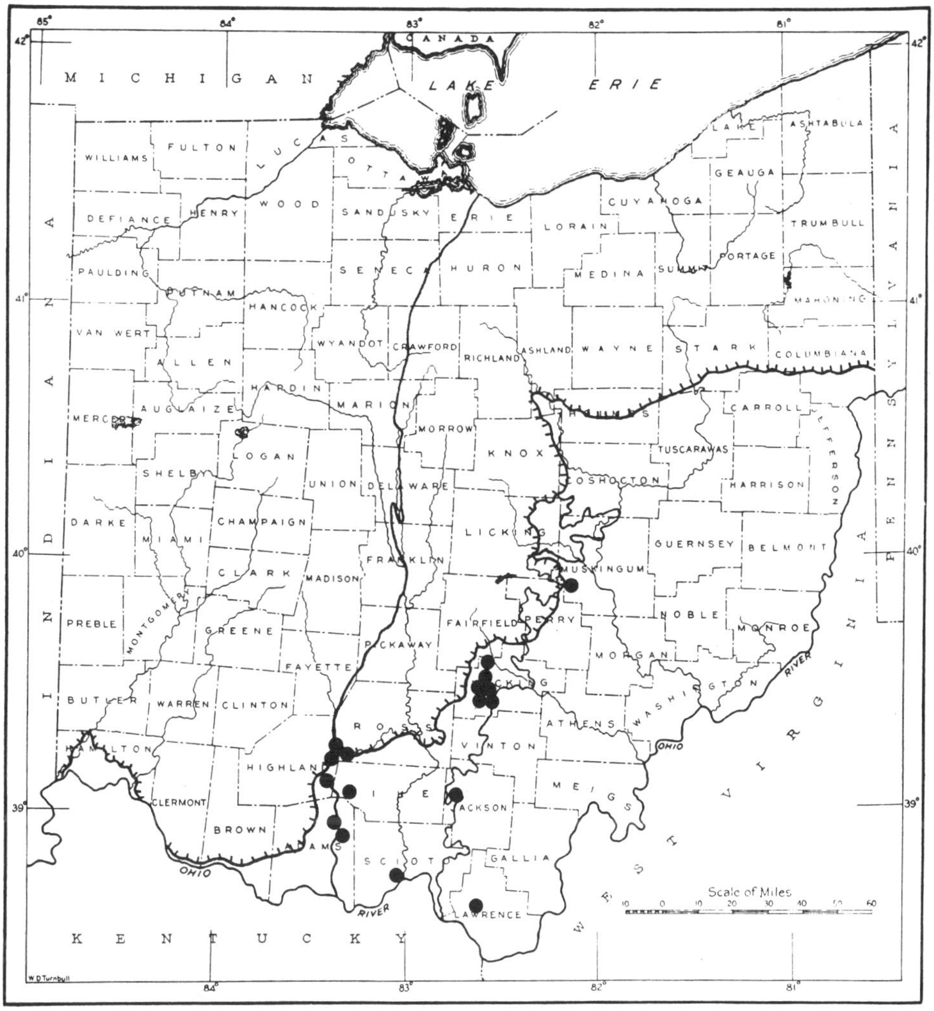

Fig. 10. Map of Ohio showing distribution of *SULLIVANTIA SULLIVANTII* based on herbarium specimens.

COLLECTIONS OF 1840 AND THEIR FATE

The largest known collection of Sullivant's flowering plants remaining as a single unit is in the herbarium of The Ohio State University. This collection of 390 specimens is the second oldest in the herbarium and was transferred from the University's Department of Horticulture in the 1950s or 1960s (Stuckey and Wentz, 1974), although part of the collection may have been the 375 specimens received in 1940 (Jones, 1940). The specimens of the entire collection are all labeled with the name of the plant as originally identified and with the data, "Columbus, Ohio 1840 W. S. S." (Fig. 11A). From the kinds of plants in the collection and a knowledge of the present distribution of many of these species in Ohio, it is evident that not all of these plants came from Columbus and vicinity as the data on Sullivant's original labels indicate. As was the custom in his day, Sullivant wrote "Columbus,

Ohio" on the labels, because that was his home address and that information was of greater importance to those botanists with whom he was exchanging plants, rather than the actual geographical locality whence the plant came. Because of these inadequate labels, some botanists (cf. Meyer, 1951; Hauser, 1963) have inadvertently attributed certain rare or infrequent species in Ohio to Columbus or Franklin County, when in fact the plants do not, and apparently never did occur in the central Ohio area. If one considers and correctly assumes that the plants were indeed collected in the summer of 1840, then this set of specimens was not the basis for Sullivant's *Catalogue*, published in February of that year. Furthermore, specimens representing 77 taxa in Sullivant's collection were not included in his *Catalogue* (1840). Those specimens that are the same as those mentioned in the *Catalogue*, however, do serve as vouchers for the concepts of these taxa, and for that reason alone, this set of specimens is one of the most valuable collections of Ohio plants in the herbarium of The Ohio State University. Not only does Sullivant's collection give a documented record of the 1840 flora of the vicinity of Columbus and Franklin County, but it also contains some specimens of species that have never been rediscovered in the area and even some which are little known, rare, or have now disappeared from the state. A manuscript list of 436 taxa has been compiled from Sullivant's specimens in The Ohio State University Herbarium, in addition to specimens seen in other herbaria (Roberts and Stuckey, 1975).

Sullivant's activity for the plant collecting season of 1840 is recorded in letters he wrote to Asa Gray, John Torrey, and Charles W. Short, and the information from them further supports the hypotheses that this collection of plants was made in 1840 and that some of the plants came from localities in Ohio other than Franklin County. On 26 April 1840, Sullivant wrote to Torrey:

> It is my intention to visit this season many promising localities for plants within our state.

On 27 May to Gray he said,

> In these dull times instead of turning my attention to politics as everybody does, I attend almost entirely to making short botanical trips around the country of 2, 3, or 4 days . . . .

He described his most extensive trip of the season to Gray in a letter of 31 July:

> My . . . absence of near three weeks . . . was on a trip undertaken to examine a portion of our state that has been (as far as I can learn) but little visited by botanists; the high table lands dividing the water of Lake Erie from those of the Ohio River, back of Cleaveland & easterly towards the Pennsylvania line. These lands are considerably elevated, and abound in small lakes and swamps interspersed with the Tamarack (*Pinus pendula*) [*Larix laricina*] and other coniferae [probably *Tsuga canadensis*]. -- But unluckily I failed in my object owing to a slight sickness, not severe but sufficient to prevent active exertion. I spent most of my time at Cleaveland & the Falls around which places I made a few collections -- I was however able for one day to examine the Summit Lake that feeds the Akron summit of our canal. This was a very interesting place -- I there found several plants of the rarer kind, particularly aquatics [*Myriophyllum heterophyllum* (specimen at PH)], that I never before met with. -- Of these and my other collections this season, I will report to you more fully at an other time. --

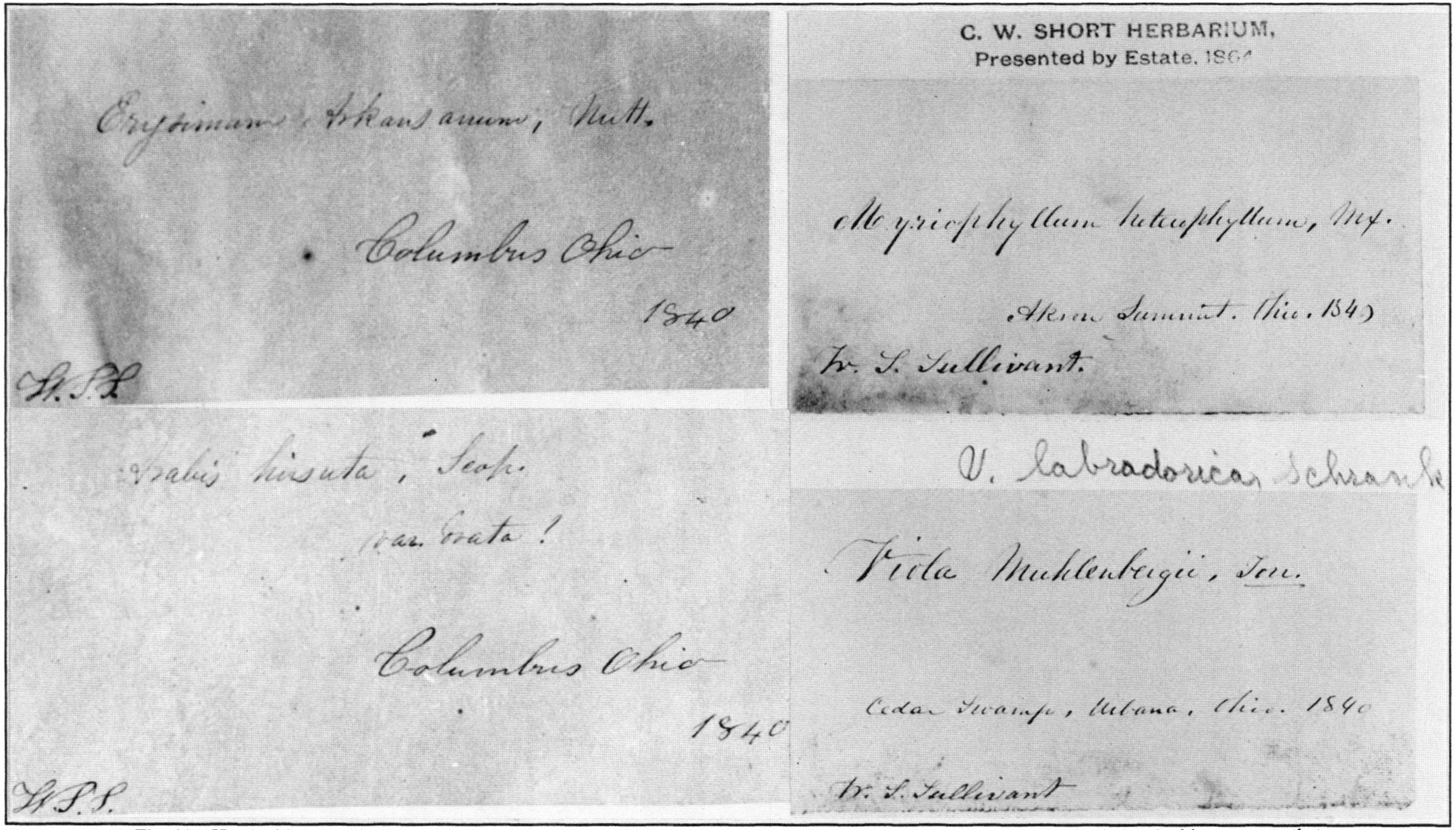

Fig. 11. Handwriting on selected labels of plants collected by William S. Sullivant. The neat uniform appearance of this handwriting suggests that these labels may have been prepared by his wife, Eliza G. W. Sullivant. The labels illustrated are from specimens of: A. *Erysimum arkansanum* (OS) and *Arabis hirsuta* (OS); **B.** *Myriophyllum heterophyllum* (PH) and *Viola muhlenbergii* (PH).

He offered to send Torrey and Gray plants, as mentioned further in the letter.

> You say you are in want of N. American plants for your European friends, I am glad that I can furnish you with a good many of our Ohio plants -- I shall send you in the fall a large supply & any thing you may have to spare European or North American will be acceptable. -- I wish to increase my N[orth] A[merican] herbarium & should you or Dr. Torrey come across any of our plants other than those common to this part of the country, for exchange or sale at any thing like a reasonable price, please secure them for me -- exchanges I find tedious & uncertain for one living in the interior -- Should you have any plants to lay by for me, put in a good portion of Cyperaceae & Gramineae, with [the publication by] Kunth I am pretty prepared for [studying] those orders.

In similar words, Sullivant also told of his trip to northeastern Ohio in a letter of 27 August 1840 to Charles W. Short of Louisville, Kentucky. He further noted that, "Thus far, this season, I have not been as successful in collecting additional species I anticipated." He offered to send Dr. Short plants, the names of which could be selected from his 1840 *Catalogue* and from an appended list of 72 plants given in the letter. The latter plants Sullivant specifically stated he "collected this season (1840) . . . ." In this list (Appendix VI), Sullivant marked 24 of them with a short line beside the name. Most of these grow in Ohio only along the sandy shore of Lake Erie or in the tamarack bogs of northeastern Ohio. Specimens representing 22 of the 72 names are in the Sullivant collection (at OS).

Sullivant distributed his plants widely, and many of them are now in other herbaria. Previous to 1840, most of Sullivant's specimens of flowering plants that were sent to John Torrey (NY) and Asa Gray (GH) were not dated, as are many of those seen in the George Engelmann or Samuel B. Buckley herbaria (both at MO). Sullivant also sent specimens to Charles W. Short (Fig. 11B), some of which remained in his herbarium and upon his death were presented by his estate in 1864 to the Academy of Natural Sciences of Philadelphia (PH). Many of these specimens do have locality data other than Columbus, Ohio. Information in letters to Torrey (1 September 1840, 17 October 1840, -- July 1841) and to Gray (18 August 1841) reveals that Sullivant sent specimens of his 1840 collections to these botanists, and some of them are now in the herbaria of the New York Botanical Garden (NY) and the Gray Herbarium of Harvard University (GH), respectively. Other specimens evidently were intended for European botanists, as Sullivant noted in his 18 August 1841 letter to Gray,

> You and Dr. Torrey have misunderstood me about the four large bundles of Ph[an]erog[amous] plants. I sent them as presents to you both, in consequence of a remark from you that plants would be acceptable for your foreign correspondents--not to be exchanged for my benefit--I consider them very inadequate returns for the many valuable plants I have received from you & him.

The known localities in Ohio from which Sullivant obtained flowering plants are marked on a map (Fig. 12). A list of specimens known to have been procured from these sites has been prepared from various sources (Appendix VII).

OTHER CONTRIBUTIONS DURING THE 1840S

As early as 1842, Sullivant began sending Gray living specimens of flowering plants to be planted in the Cambridge Botanic Garden of Harvard University or used to illustrate some of his books. Some of the earliest living specimens sent were his namesake, *Sullivantia ohionis* (Rodgers, 1940, p. 133). Gray hired Isaac Sprague as his botanical artist to prepare the illustrations. Sprague was the major botanical illustrator in the United States at this time. He provided accurate and artistically pleasing drawings for both John Torrey and Asa Gray, and thus aided them in becoming accepted as world-class botanists (Rudolph, 1990). A large full-size page drawing of *Sullivantia ohionis* A. Gray [= *S. sullivantii* Torrey & Gray] was included in his *Chloris Boreali-Americana* (Gray, 1848). In the second volume of the *Genera Florae Americae Boreali-Orientalis Illustrata* (Gray, 1849), Isaac Sprague illustrated eight species based on living specimens that Sullivant obtained in Ohio for Gray. These were sent to Gray on two occasions during the summer of 1846. The plants were *Isopyrum biternatum* Torr. & Gray (false rue anemone), *Delphinium tricorne* Michx. (dwarf larkspur), *Asimina triloba* Dunal (papaw), *Stylophorum diphyllum* Nutt. (celandine poppy), *Iodanthus*

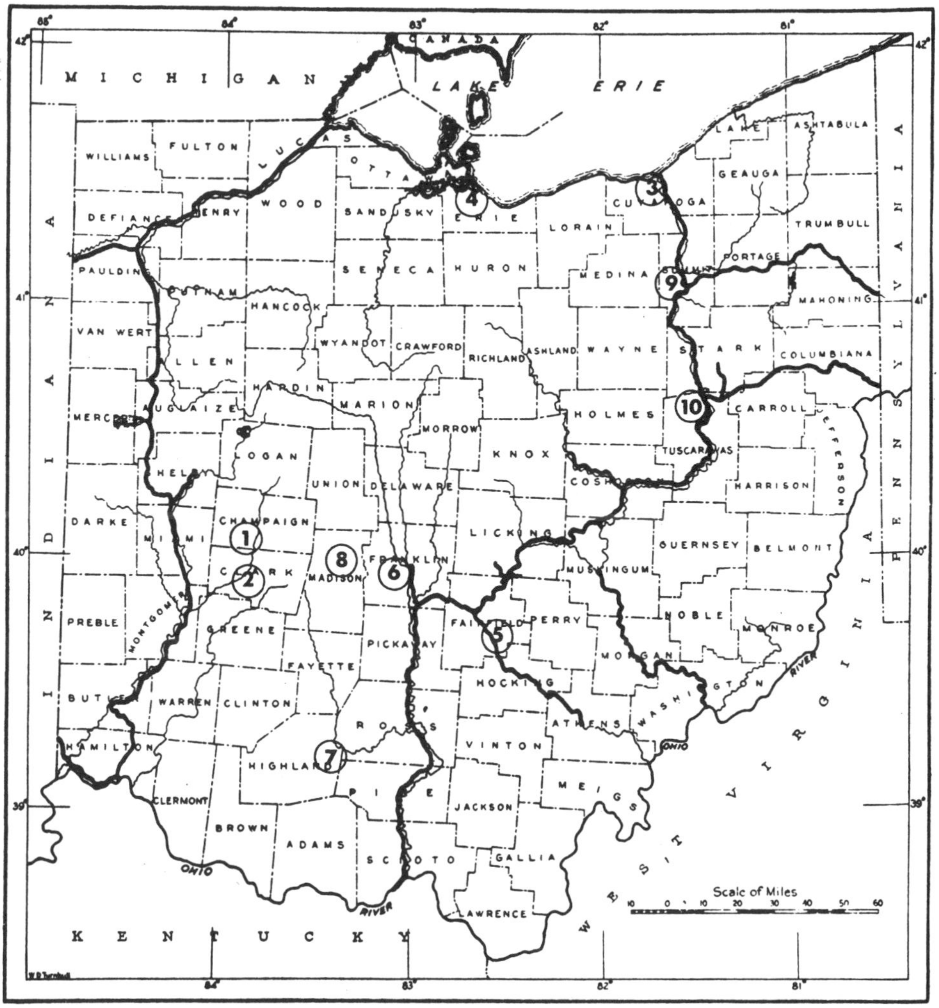

Fig. 12. Map of Ohio showing William S. Sullivant's known flowering plant collecting localities. By assembling all of Sullivant's data that indicate a place where a plant was obtained or seen, it has been possible to compile a list of 10 Ohio counties where he botanized (Appendix VII). These data have been obtained from Sullivant's CATALOGUE (1840); from his "Account of three undescribed plants of central Ohio" (1842); from an unpublished version of the latter article filed with his correspondence to Asa Gray (at GH); from letters to Gray (at GH), John Torrey (at NY), and Charles W. Short (at the Filson Club); and from labels on specimens at GH, MO, NY, and PH. The solid black lines on the map show the routes of the canals in Ohio. Sullivant traveled extensively by canals when botanizing.

hesperidoides Torr. & Gray (false rocket), *Arabis patens* Sulliv. (spreading rock-cress), *Erysimum arkansanum* Nutt. (western wallflower) and *Napaea dioica* L. (glade mallow).

Sullivant's last significant contribution to phanerogamic botany was his preparation for publication of Thomas G. Lea's *Catalogue* (1849) of plants of the vicinity of Cincinnati, Ohio. Lea, also a former businessman, had studied in detail the flora of the Cincinnati area for 10 years, and prior to his death in 1844, expressed in writing his desire that Sullivant or John Carey prepare his manuscript for publication (Rodgers, 1940, pp. 166-167). Lea's son, James M. Lea, later remarked that he was indebted to Sullivant ". . . for the very complete and perfect form . . ." that resulted in the published *Catalogue* (Rodgers, 1940, p. 187). Lea's local flora was the first published for the Cincinnati area (revised by Braun, 1934).

WANING INTEREST IN PHANEROGAMIC BOTANY AND DISPERSION OF HERBARIUM

As documented in letters to Asa Gray (6 January 1840 and 9 November 1840), Sullivant was then beginning his intensive studies of the mosses and the liverworts, which expanded considerably over the next several years. By 1845 his enthusiasm for phanerogamic botany had begun to wane, and this change is revealed in four letters.

To Asa Gray (19 August 1845) he wrote:

> In fact--the only way I expect to keep alive my fading away phanerogamic Botany is by reading your new Editions of Text Book--hope (you the same?) the intervals will be short--these Musci Alleghen[iensis] have so completely buried me in such matters I fear I shall never get my nose out of them again.

To George Engelmann (18 September 1849), he revealed:

> . . . Phanerogamous botany I have long since ceased to study critically -- my occupations permit me to keep the study only of mosses and Hepaticae hence I am unable to meet satisfactorily some of your querries.

Later, to Edward Tatnall (19 November 1860), published in Harshberger (1899, p. 226), Sullivant said,

> Thanks for the two Sagittarias. Pha[n]erogamous botany, though I am sorry to say . . . with me is divided into two parts--one that I knew but have forgotten, and the other I never knew.

To Harry N. Patterson (21 March 1866), published in Kibbe (1953, p. 450), he commented,

> For the last two years I have ceased making collections of Ph[an]e[r]ogamous plants entirely, except a collection I made in the Alleghenies last summer with Dr. Gray . . . .

By 1846 Sullivant had decided to disperse the remainder of his flowering-plant herbarium. To Gray (12 July 1846), he said:

> The first spare day I can find I shall pack up & send to you all my spare specimens of Ph[an]erogams out of which [John] Carey must take what he wants, the remainder you can put to better use than I can.

Earlier, in a letter of 8 November 1845, Sullivant indicated to Gray that he would aid in the restoration of John Carey's herbarium, which had been destroyed by fire. Later, in 1846 (12 September), Sullivant wrote to Gray:

> The two boxes mentioned in my note the other day are on the way for you via canal -- One contains 7 or 8 bundles of Phanerogamous plants, my entire stock of duplicates, Mostly Ohio coll[ections] including the Alleghany Coll[ection] from which last I took only a set for my Herb. From the above I wish Mr. Carey to supply his Herb. with what he may want--the remainder are yours they may be of some use to you in foreign exchanges.

Still later, Sullivant told what he did with his personal herbarium that he had retained after sending the duplicates to Gray in 1846. On 11 January 1868, he revealed to Gray:

> About my Pha[n]erogamous Herbarium -- that and much of my cryptogams and many books are boxed up (15 to 20 boxes) in the corn crib and in such a condition I cannot now get at them. In the spring I shall shelve off a room for them where I can see what is best to be done with them.

That spring, he made the final disposition of his phanerogamic herbarium, as written to Gray on 22 April:

> Dear Gray -- I have lately unboxed my botanical things and have concluded (following the suggestion in your last letter) to send you all my Pha[n]erogamous plants--accordingly, I have this day sent by freight lines three boxes
>
> No. 1 -- Plants not North America
>
> No. 2 -- North Amer[ican] plants only (both arranged according to Amer. Ed. of Lindley)
>
> No. 3 -- Bundles unarranged among them, the Fendler & Lindheimer Texan collections, &c &c.
>
> In all there must be at least 600 species and 10,000 specimens. Do what you please with them, they have served their day with me--and a very pleasant one it has been. -- After all there is a little touch of melancholy about the matter. It looks like one administering on his own estate--in my case beginning with the botanical assets. -- I have given evry thing I had in the Hepatical way, to [Coe Finch] Austin. -- shall retain nothing but the Musci which will be quite sufficient for any time I may have for such things.

The fine herbarium praised by Alva Curtis in his reviews of the Sullivant brothers' *Catalogues* had now all left Columbus to be disposed of as Asa Gray desired. With this final transaction completed, the question remains of the source of the extensive collection in the herbarium of The Ohio State University. Rodgers (1940, p. 255) noted that Sullivant willed his remaining collections, not given to Harvard University, to The Ohio State University. This statement seems contradictory because, as noted above, the flowering plants supposedly had earlier all been given to Asa Gray. This collection may not have been one retained personally by William S. Sullivant, but one given at an earlier date to his brother Joseph, who was instrumental in the founding of the University (Kinnison, 1970). Joseph could have left it to the University in the Department of Horticulture and Botany, which upon its separation in 1891 formed the Department of Horticulture and Forestry and the Department of Botany (Rudolph and Stuckey, 1969; Meyer, 1983). The Sullivant collection evidently was retained with the Department of Horticulture, where Joseph's main scientific interest lay, and was not returned to the Department of Botany until the mid-twentieth century (Stuckey and Wentz, 1974).

CONTRIBUTIONS AND EVALUATION

While studying the phanerogamic flora of Ohio in the late 1830s and early 1840s, Sullivant's botanical achievements as a "neophyte" were: (1) his extensive travels in Ohio and his correspondence with fellow botanists, (2) his acquisition of a large herbarium of flowering plants with special emphasis on grasses and sedges, (3) his published *Catalogue of Plants . . . in the Vicinity of Columbus, Ohio* (1840), and the preparation for publication of Thomas G. Lea's *Catalogue of Plants . . . in the Vicinity of Cincinnati, Ohio . . .* (1849), (4) his keen observations on variations and hybridization among certain species of plants, particularly *Carex*, and (5) his discovery and publication of plants new to science. With reference to the last item, the four species that Sullivant described as new to science are recognized today as valid entities of the flora, and most of the taxa described from his specimens by other botanists are also retained today. These discoveries are an indication that Sullivant's early botanical work was based on careful field observations and critical microscopic study leading to sound, accurate, and lasting scientific work.

An evaluation of Sullivant as a botanist at the height of his work in phanerogamic botany comes from Dr. Charles Wilkins Short, well respected and widely known botanical collector and correspondent of Louisville, Kentucky (Stuckey, 1978a, b; Skaggs, 1982). In response to Asa Gray's question asking for names of individuals from the West and South who could supply plants and seeds for the Cambridge Botanic Garden, Short in a letter of 26 December 1842 listed several names, the last one being Sullivant, of whom Short wrote further:

> The last named individual is a man of fortune -- a rara avis among naturalists -- and backs his ability by a most ready and obliging willingness to render all aid in the advancement of our common cause. Mr. Sullivant possesses other gratifications as rarely united, I believe, to wealth, as wealth is to the pursuit of Science, -- he is indefatigably industrious and labourious; so that I think you may expect more from him than from any other Western Naturalist.

Sullivant's physician-botanist friend from Lancaster, Ohio, John M. Bigelow (1843), noted that he was an "ardent, intelligent, and scientific cultivator of this interesting branch of natural history." Sullivant and his wife first became "amateur botanists of a few acres, [then] they extended their botanical researches, not only over Ohio, but in a measure over the whole known world. In the short space of time in which he has been engaged in these pursuits, Mr. Sullivant has, probably, personally collected more plants than any other botanist of Ohio."

In a letter dated 6 December 1857 to Sir William Hooker, Director of the Kew Gardens in England, Asa Gray wrote: "A noble fellow is Sullivant and deserves all you say of him and his works. The more you get to know of him the better you will like him." It was from this passage that Rodgers (1940) chose the title "Noble Fellow" for his biography of Sullivant.

In his obituary of Sullivant, Gray (1873) praised him for laying a broad and complete foundation for the study of bryology in this country. Gray further wrote that Sullivant's works always will be the classic authority and for a long time will be unrivalled.

> In personal appearance and carriage, no less than in all traits of an unselfish and well-balanced character, Mr. Sullivant was a fine specimen of a man. He had excellent business talents, and was an exemplary citizen; he had a refined and sure taste, and was an accomplished draughtsman . . . As an investigator he worked deliberately, slowly indeed and not continuously, but perseveringly."

Among the honors he received were: election to the National Academy of Science in 1872, an honorary degree of Doctor of Laws from Kenyon College, Ohio, and memberships in the principal scientific societies in America and Europe. His name is commemorated in numerous species of plants, including fungi, liverworts, mosses, flowering plants, and a fossil plant (Armstrong, 1901; Wareham, 1940). Gray (1873) further noted that Sullivant's

> . . . oldest botanical associates . . . enjoyed the pleasure of bestowing the name *Sullivantia ohionis* [= *S. sullivantii*] upon a very rare and interesting, but modest and neat Saxifragaceous plant, which he himself discovered in his native State, on the secluded banks of [Rocky Fork Creek], a tributary of the [Scioto] river which flows by the place where he was born, and where his remains now repose [then in the old Franklinton Cemetery, since removed to Green Lawn Cemetery, Columbus (Stuckey, 1987b)].

In Germany, Müller (1875) praised Sullivant as an American who met European standards of science in the mid-nineteenth century. He also commented that Sullivant's wife took pleasure in his scientific studies, and aided him extensively in the preparation of his initial works on the mosses.

Thomas C. Mendenhall (1915), Professor of Physics and later a member of the Board of Trustees of The Ohio State University, wrote that "eminent men of science from different parts of the world were frequently guests" at the Sullivant home. Mendenhall also had visited in Sullivant's home many years previously, and characterized him as

> . . . an admirable . . . scientific man . . . , [a] man of genius and scholarly instincts, who, free from the restriction upon his intellectual liberty, pursues with ardent zeal the investigation of some department of science finding his reward in the unalloyed delight of original discovery.

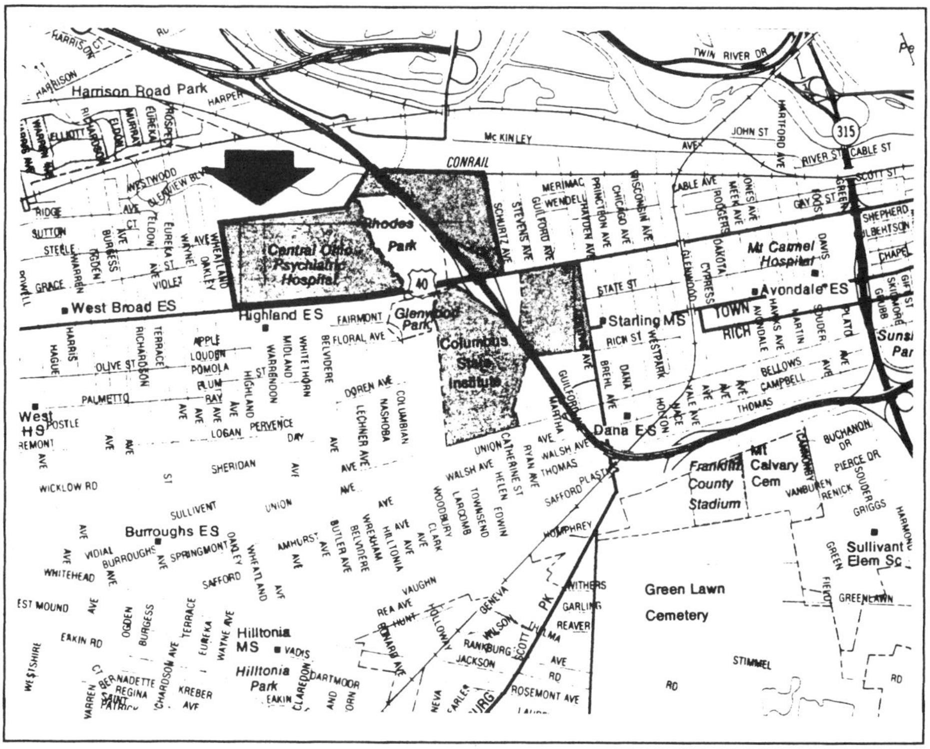

Map of Columbus showing location of Central Ohio Psychiatric Hospital and Rhodes Park, formerly "Sullivant Hill."

EPILOGUE

William S. Sullivant (Fig. 13) was thrice married to very young ladies (Figs. 14A-C) who had not yet reached the age of 20. His first marriage was 7 April 1824 to Jane Marshall (1808-1825), daughter of Alexander Keith Marshall of Mason County, Kentucky, and niece of John Marshall, the distinguished Chief Justice of the United States. To them was born Jane Marshall, their only child, as Mrs. Sullivant died shortly after the child's birth. His second marriage was 29 November 1834 to Eliza Griscom Wheeler (1817-1850) of New York City, with whom he had Lucas, Eliza G. (who married Andrew Denny Rodgers I), William Starling, Edmund Gray, Henry, and Emma. A little over a year after Eliza's death from cholera, he married Caroline Eudora Sutton (1833-1891) of Aurelius, Cayuga County, New York, on 1 September 1851. To this union were born Caroline Louisa (Kitty), Thomas Starling, Jacob Sutton, Allen, Edith Steele, and Roderic Gordon (Fig. 15). In January 1873, William contracted pneumonia and became bedridden. His condition grew much weaker until his death in April of the same year. William S. Sullivant is buried in the Sullivant family plot in Green Lawn Cemetery, Columbus, with his three wives, all four sons from his second marriage, none of whom lived past 24 years (Fig. 16), and Edith and Roderic from his third marriage.

From his tombstone (Fig. 17) in Green Lawn Cemetery, Columbus, Ohio:

William Starling
Sullivant, L.L.D.
A Classical Scholar
and
Member of Various Learned
Societies in Europe and America

Born Jan. 15, 1803,
Died April 30, 1873.

SULLIVANT

William S. Sullivant family plot in Green Lawn Cemetery. Tombstones left to right: children, Henry and Lucas; wife Eliza G.; William S.; wife Caroline S.; wife Jane M. (photograph by Ronald L. Stuckey, November 1990).

Fig. 13. William S. Sullivant (drawn from a photograph in Rodgers, 1940, opp. p. 226; original in the Herbarium of The Ohio State University.)

Fig. 14 A. First of the three wives of William S. Sullivant: Jane Marshall Sullivant (from Rodgers, 1940, opp. p. 86).

Fig. 14 B. Second of the three wives of William S. Sullivant: Eliza Griscom Wheeler Sullivant (from the Ohio Historical Society, Columbus).

Fig. 14 C. Third of the three wives of William S. Sullivant: Caroline Eudora Sutton Sullivant (from Rodgers, 1940, opp. p. 212).

Fig. 15. Three of the four sons of William S. and Caroline E. S. Sullivant. l. to r. Roderic (1863-1904), Allen (1860- x), and Jacob (1858-1918) (from the Ohio Historical Society, Columbus).

Fig. 16. The two monuments marking the gravesites of the four sons of William S. and Eliza G. W. Sullivant. Front marker: Henry (1846-1849) and Lucas (1836-1838); back marker: William Starling (1841-1865) and Edmund Gray (1844-1867) (photograph by Melanie Pratt, December 1986).

Fig. 17. Monument of William S. Sullivant (photograph by Ronald L. Stuckey, November 1990).

REFERENCES CITED

[ANONYMOUS]. 1845. Columbus Horticultural Society. Ohio Cultivator 1: 77.

[____]. 1878a. Sudden death [of Dr. Jonathan R. Paddock]. Maysville (Kentucky) Republican, 8 June.

[____]. 1878b. Death of Dr. J. R. Paddock. Maysville (Kentucky) Bulletin, 13 June.

[____]. 1894. Asa Horr, pp. 135-136. *In* Portrait and Biographical Record of Dubuque, Jones and Clayton Counties, Iowa. Chapman Publishing Co., Chicago.

[____]. 1901. [Lucas Sullivant], pp. 13-14. *In* A Centennial Biographical History of the City of Columbus and Franklin County, Ohio. Lewis Publishing Co., Chicago. 1012 pp. port.

[____]. 1920. [Herbarium of Jonathan R. Paddock]. Science n. ser. 51: 165.

ARMSTRONG, CLARA. 1901. Plant names commemorative of Ohio botanists. Ohio Naturalist 1: 33-38.

ATKINSON, WILLIAM B., ed. 1878. Asa Horr, p. 237; John Milton Bigelow, p. 285. *In* The Physicians and Surgeons of the United States. Charles Robson, Philadelphia. 788 pp.

BIGELOW, JOHN M. 1841. *Florula Lancastriensis*: Or a catalogue, comprising nearly all the flowering and filicoid plants, growing naturally within the limits of Fairfield County, with notes of such as are medicinal. Proc. Medical Convention of Ohio, 5-7 May 1841: 49-79.

____. 1843. On the topography, etc. of Fairfield County, O[hio]. Western Lancet 2: 345-360.

____. 1849. [A list of all the medicinal plants of Ohio (not embraced in Wood and Bache's Dispensatory of the United States), containing as far as known, a brief account of their properties]. Ohio Med. & Surg. J. 2: 97-143. Reprinted, 1849. J. J. Riley & Co., Columbus. 47 pp.

____ and ASA HORR. 1841. *Florula Lancastriensis*: Or a Catalogue Comprising Nearly All the Flowering and Filicoid Plants, Growing Naturally within the Limits of Fairfield County. Lancaster, Ohio. 22 pp. [With the list of medicinal plants and accompanying notes omitted.]

BOOTT, FRANCIS. 1842. *Carex sullivantii*, p. 29. *In* Asa Gray. Notes of a botanical excursion to the mountains of North Carolina. Amer. J. Sci. Arts 42: 1-49.

____. 1862. Illustrations of the Genus *Carex*. Part Third. William Pamplin, London. iv, 105-206 + tab. 311-411. (Facsimile ed., 1968. J. Cramer, Lehre, West Germany.)

BOYNTON, C. L., and C. D. BEADLE. 1901. Revision of certain sunflowers. Biltmore Bot. Stud. 1(1): 11-18.

BRAUN, E. LUCY. 1934. The Lea Herbarium and the flora of Cincinnati. Amer. Midl. Naturalist 15: 1-75.

____. 1961. The Woody Plants of Ohio: Trees, Shrubs, and Woody Climbers Native, Naturalized, and Escaped. Ohio State University Press, Columbus. 362 pp. (Facsimile ed., 1969. Hafner Publishing Co., New York.)

____. 1967. The Monocotyledoneae [of Ohio]: Cat-tails to Orchids. With the Gramineae by Clara G. Weishaupt. Ohio State University Press, Columbus. ix, 464 pp.

BRITTON, N. L. 1913. Saxifragaceae, pp. 177-180. *In* List of Pteridophyta and Spermatophyta growing without cultivation in northeastern North America. Mem. Torrey Bot. Club 5: 1-377.

BURR, C. B., ed. 1930. [John M. Bigelow], pp. 248-249. *In* Medical History of Michigan, Vol. 2. Bruce Publishing Co., Minneapolis.

CHUANG, TSAN-LANG, and LINCOLN CONSTANCE. 1969. A systematic study of *Perideridia* (Umbelliferae-Apioideae). Univ. California Publ. Bot. 55: 1-74.

CROSBY, JAMES O. 1915. Scientific studies of Dr. Asa Horr. Annals of Iowa ser. 3, 12: 161-164 + port. opp. p. 161.

[CURTIS, ALVA]. 1839. [Review of] An alphabetical catalogue of shells, fossils, minerals, and zoophites, in the cabinet of Joseph Sullivant. Botanico-Medical Recorder 7: 219.

[____]. 1840. [Review of] A catalogue of plants, native or naturalized in the vicinity of Columbus, Ohio--By Wm. S. Sullivant. Botanico-Medical Recorder 8: 188-189.

[DENNY, GUY L.]. 1987. Another exciting find [*Nothoscordum bivalve*]. Newsletter, Ohio Dept. Nat. Resources, Div. Nat. Areas & Preserves 9(4): [6].

DEWEY, CHESTER. 1824-1835. Caricography. Amer. J. Sci. Arts 7: 264-278; 8: 93-99, 264-267; 9:60-73, 257-263; 10: 30-48 + pl., 265-284; 11: 147-167, 304-325; 12: 296-297; 14: 351-354 + 2 pl.; 25: 140-146 + pl.; 26: 107-108, 376-378; 27: 236-241 + pl.; 28: 270-276.

____. 1845. Caricography. Amer. J. Sci. Arts 49: 42-48. [*Carex sullivantii*, p. 44, pl. Ee, fig. III.]

D[RAKE, DANIEL]. 1840. [Review of] A catalogue of plants, native and naturalized, in the vicinity of Columbus, Ohio. By Wm. S. Sullivant, 1840, p. 63. West. J. Med. & Surgery 1: 278-280.

DODDS, GILBERT F. 1959. Early Horticulture in Franklin County. Franklin County Historical Society and Museums, Special Bulletin No. 4. 15 pp.

EASTERLY, NATHAN W. 1964. Distribution patterns of Ohio Cruciferae. Castanea 29: 164-173.

EGGERS-WARE, DONNA MARIE. 1970. A revision of *Valerianella* in North America. Ph.D. dissertation, Vanderbilt University, Nashville, Tennessee. 262 pp.

____. 1983. Genetic fruit polymorphism in North American *Valerianella* (Valerianaceae) and its taxonomic implications. Syst. Bot. 8: 33-44.

ENGELMANN, GEORGE. 1848. *Asclepias sullivantii*, p. 366. *In* Asa Gray. A Manual of Botany of the Northern United States . . . Arranged According to the Natural System. James Munroe & Co., Boston and Cambridge. lxxii, 710 pp.

____. 1866. A revision of the North American species of the genus *Juncus* with a description of new or imperfectly known species. Trans. St. Louis Acad. Sci. 2: 244-498.

FELTER, HARVEY WICKES. 1902. History of the Eclectic Medical Institute Cincinnati, Ohio, 1845-1902, Including the Worthington Medical College (1830-1842), the Reformed Medical School of Cincinnati (1842-1845), and the Eclectic College of Medicine (1856-1859), with Biographical Sketches of Members of the Various Faculties and Lists of Graduates Arranged Alphabetically and by Classes. Published for the Alumni Association of the Eclectic Medical Institute by H. W. Felter, M.D., John K. Scudder, M.D., and J. U. Lloyd, Phr. M., Committee, Cincinnati, Ohio. 203 pp.

FERNALD, MERRITT LYNDON. 1950. Gray's Manual of Botany. Ed. 8. American Book Co., New York. lxiv, 1632 pp. (Reprinted, 1970. D. Van Nostrand Co., New York.)

GRAY, ASA. 1842. Notes of a botanical excursion to the mountains of North Carolina. Amer. J. Sci. Arts 42: 1-49. [*Sullivantia ohionis*, pp. 21-22, 29.]

____. 1848. Chloris Boreali-Americana: Illustrations of New, Rare, or Otherwise Interesting North American Plants Selected Chiefly from Those Recently Brought into Cultivation at the Botanic Garden of Harvard University[,] Cambridge. Decade I. Mem. Amer. Acad. Arts n. ser. 3: 1-56 + 10 pl. [*Sullivantia ohionis* Torrey and Gray, pl. 6].

____. 1849. Genera Florae Americae Boreali-Orientalis Illustrata. The Genera of the Plants of the United States Illustrated by Figures and Analyses from Nature, by Isaac Sprague, Vol. 2. George P. Putnam, New York and London. 230 pp. + pl. 101-186.

____. 1857. Letter to William J. Hooker, 6 December, from Cambridge, Massachusetts. Printed, 1893. Pg. 433. *In* Jane Loring Gray, ed. Letters of Asa Gray, Vol. 2. Houghton, Mifflin and Company, Boston. iv, 369-837 pp.

____. 1873. William S. Sullivant: A biographical notice. Amer. J. Sci. Arts 106: 1-6.

____. 1874. Biographical sketch [of William Starling Sullivant], pp. 1-8. *In* William S. Sullivant. Icones Muscorum. Charles W. Sever, Cambridge, Massachusetts. 109 pp. (Facsimile ed., 1969. S. H. Service Agency, New York.)

____. 1877. William S. Sullivant. Biog. Mem. Natl. Acad. Sci. 1: 277-285.

____. 1883. Contributions to North American botany. II. Miscellaneous genera and species [*Lonicera*]. Proc. Amer. Acad. Arts & Sci. 19: 74-96.

HARSHBERGER, JOHN W. 1899. The Botanists of Philadelphia and Their Work. T. C. Davis & Sons, Philadelphia. xii, 457 pp.

HAUSER, EDWARD J. P. 1963. The Dipsacaceae and Valerianaceae of Ohio. Ohio J. Sci. 63: 26-30.

HEISER, CHARLES B., JR. 1969. The North American sunflowers (*Helianthus*). Mem. Torrey Bot. Club 22(3): [i-ii], 1-218.

HOPKINS, MILTON. 1937. *Arabis* in eastern and central North America. Rhodora 39: 63-98, 106-148, 155-186 + pls. 457-458. (Reprinted, 1937. Contr. Gray Herb. 116.)

HOWE, E. C. 1881. *Carex sullivantii*, Boott., a hybrid. Bot. Gaz. 6: 169-170.

JACKSON, R. C., and A. T. GUARD. 1957. Natural and artificial hybridization between *Helianthus mollis* and *H. occidentalis*. Amer. Midl. Naturalist 58: 422-433.

JEPSON, WILLIS LINN. 1962. California botanical explorers -- XII. John Milton Bigelow. Madrono 16: 179-183.

JONES, CLYDE H. 1940. Additions to the revised catalogue of Ohio vascular plants. VIII. Ohio J. Sci. 40: 200-216.

JONES, EMMA, ed. 1913. [Letter to Dr. Starling Loving, 15 August 1907], pp. 112-114. *In* Correspondence of the Late James Kilbourne . . . Published by the editor, Columbus, Ohio. 118 pp. + addenda.

KELLY, HOWARD A. 1928. Asa Horr (1817-1896), pp. 597-598. *In* Howard A. Kelly and Walter L. Burrage, eds. Dictionary of American Medical Biography; Lives of Eminent Physicians of the United States and Canada from the Earliest Times. D. Appleton and Co., New York. xxx, 1364 pp. Reprinted, 1971. Milford House, Boston.

KIBBE, ALICE L., ed. 1953. Afield with Plant Lovers and Collectors. Botanical Correspondence of the Late Henry N. Patterson with the Great Botanical Collectors and Distributors of America from 1870 to 1919. Published by the author, Carthage College, Carthage, Illinois. 565 pp.

KING, CHARLES C. 1981. In support of Eliza Sullivant as the discoverer of *Sullivantia sullivantii* [abstract]. Ohio J. Sci. 81 (April Program Abstracts): 19.

KINNISON, WILLIAM A. 1970. Building Sullivant's Pyramid: An Administrative History of the Ohio State University, 1870-1907. Ohio State University Press, Columbus. xii, 225 pp.

LAPHAM, INCREASE ALLEN. 1811-1875. Letters, diaries, and papers. Compiled by his daughter Julia Lapham. Library, Historical Society of Wisconsin. Also in typescript, Lapham Papers (MS Coll. 216), Library-Archives Division, Ohio Historical Society, Columbus.

LAZENBY, W. R. 1892. Comparison of the Sullivant Catalogue of the plants of Franklin County, Ohio, with the list of 1891 of Selby and Craig. Bull. Torrey Bot. Club 19: 16.

LEA, THOMAS G. 1849. Catalogue of Plants, Native and Naturalized, Collected in the Vicinity of Cincinnati, Ohio, during the Years 1834-1844. [Published by Isaac Lea, Philadelphia]. 77 pp.

LESLEY, J. P. 1895. Memoir of Leo Lesquereux. 1806-1889 . . . Read before the National Academy, April 16, 1890. Biog. Mem. Natl. Acad. Sci. 3: 187-212.

LESQUEREUX, LEO, and THOMAS P. JAMES. 1884. Manual of the Mosses of North America . . . With Six Plates Illustrating the Genera. S. E. Cassino and Company, Boston. vi, 447 pp. + 6 pl. Reprinted, 1895. Bradlee Whidden, Boston.

LOVING, STARLING. 1904. Some reminiscences -- general and medical. Old Northwest Geneal. Quart. 7: 217-227. Reprinted, 1934. Pp. 82-90. *In* [Harlow Lindley, ed.]. The Ohio State University College of Medicine: A Collection of Source Material Covering a Century of Medical Progress 1834-1934. Brown Publishing Co., Blanchester, Ohio. xx, 572 pp.

____. 1912. Ichabod Gibson Jones (1807-1857), pp. 52-53. *In* Howard A. Kelly, ed. Cyclopedia of American Medical Biography Comprising the Lives of Eminent Deceased Physicians and Surgeons from 1610 to 1910, Vol. 2. W. B. Saunders Company, Philadelphia and London. 545 pp. Reprinted, 1920. Pg. 637. *In* Howard A. Kelly and

Walter L. Burrage, eds. American Medical Biographies. Norman, Remington Company, Baltimore. 1320 pp. Reprinted, 1928. Pg. 673. *In* Howard A. Kelly and Walter L. Burrage, eds. Dictionary of American Medical Biography; Lives of Eminent Physicians of the United States and Canada, from the Earliest Times. D. Appleton and Co., New York. xxx, 1364 pp. Reprinted, 1971. Milford House, Boston.

MACKENZIE, KENNETH KENT. 1935. Cyperaceae (pars). Flora N. Amer. 18(5): 241-312.

MCCABE, L. R. 1887. Sketch of Leo Lesquereux. Popular Sci. Monthly 30: 835-840. port.

MEDLEY, MAX. 1980. Status report on *Sullivantia sullivantii*. *In* Endangered and Threatened Plant Status Surveys: Region IV. Kentucky. 23 pp. typescript + maps.

MENDENHALL, T. C. 1915. Some pioneers of science in Ohio. Proc. Ohio State Acad. Sci. 6: 174-203.

MEYER, BERNARD S. 1983. Botany at The Ohio State University: The first 100 years. Bull. Ohio Biol. Surv. n. ser. 6(2): i-vi, 1-177.

MEYER, FRED G. 1951. *Valeriana* in North America and the West Indies (Valerianaceae). Ann. Missouri Bot. Gard. 38: 377-503.

MÜLLER, KARL. 1875. William Starling Sullivant. Die Natur (Halle) 34: 268-270.

NOYES, J. F. 1878. [Sketch of the life of J. M. Bigelow, read at the Annual Meeting of the Detroit Academy of Medicine, 22 October]. Detroit Lancet 1: 923-926.

PORTER, THOS. C. 1872. The fedias of the northern United States. Amer. Naturalist 6: 385-388.

RIDDELL, JOHN L. 1834-1835. Synopsis of the flora of the western states. Western J. Med. & Phys. Sci. 8: 329-374. 1834; 489-556. 1835. (Reprinted as a separate, 1835. E. Deming, Cincinnati. 116 pp.)

ROBERTS, MARVIN L., and **TOM S. COOPERRIDER.** 1982. Dicotyledons, pp. 48-84. *In* Tom S. Cooperrider, ed. Endangered and Threatened Plants of Ohio. Ohio Biological Survey, Biol. Notes No. 16. 92 pp.

____, and **RONALD L. STUCKEY.** 1975. A briefly annotated revised list of W. S. Sullivant's Ohio vascular plants collected about 1840. 20 pp. typescript.

RODGERS, MRS. ANDREW DENNY. 1936. Pioneers, pp. 1-123. *In* Ruth Young White, ed. We Too Built Columbus. Stoneman Press, Columbus, Ohio. xvi, 480 pp.

RODGERS, ANDREW DENNY, III. 1940. "Noble Fellow": William Starling Sullivant. G. P. Putnam's Sons, New York. 361 pp. (Facsimile ed., 1968. Hafner Publishing Co., New York.)

____. 1942. [John Milton Bigelow], pp. 221-222, 224-226, 228, 250, 252-254. *In* John Torrey: A Story of North American Botany. Princeton Univ. Press, Princeton, New Jersey. 352 pp. Reprinted, 1965. Hafner Publishing Co., New York.

____. 1944. American Botany 1873-1892: Decades of Transition. Princeton University Press, Princeton, New Jersey. 340 pp. (Facsimile ed., 1968. Hafner Publishing Co., New York.)

____. 1962. Charles Leo Lesquereux (Nov. 18, 1806 - Oct. 25, 1889), pp. 381-382; William Starling Sullivant (Jan. 15, 1803 - April 30, 1873), pp. 612-613. *In* William Coyle, ed. Ohio Authors and Their Books: Biographical Data and Selective Bibliographies for Ohio Authors, Native and Resident, 1796-1950. World Pulishing Co., Cleveland and New York. xxiv, 741 pp.

____. 1969. Foreword, pp. [i-iv]. *In* William S. Sullivant. Icones Muscorum, or Figures and Descriptions of Most of Those Most Peculiar to North America which have not been Figured. Supplement. Posthumous. With Eighty-One Copper-Plates. S. H. Service Agency, Inc., New York. 9-109 pp. (Facsimile of the 1874 ed., Charles W. Sever, Cambridge, Massachusetts.)

ROSSBACH, GEORGE B. 1958. The genus *Erysimum* (Cruciferae) in North America north of Mexico--a key to the species and varieties. Madroño 14: 261-267.

____. 1963. Letter to Ronald L. Stuckey, 20 September, from Buckhannon, West Virginia.

RUDOLPH, EMANUEL D. 1990. Isaac Sprague, "delineator and naturalist." J. Hist. Biol. 23: 91-126.

____, and **RONALD L. STUCKEY**. 1969. History of botany in The Ohio State University. 22 pp. typescript. Archives, The Ohio State University, Columbus. Copies in the possession of the authors.

S[ARGENT], C. S. 1890. Two American honeysuckles. Garden and Forest (16 April): 187 + figs. 33 (p. 10) & 34 (p. 34).

SARTON, GEORGE. 1942. . . . Lesquereux (1806-1889). Isis 34: 97-108.

SELBY, A. D. 1899. The flora of Franklin County, Ohio [abstract]. Proc. Amer. Assoc. Advancem. Sci. 48: 300-303.

____, and **M. CRAIG**. 1890. A preliminary list of the plants of Franklin County, Ohio. J. Columbus Hort. Soc. 5: 1-19.

SHORT, CHARLES W. 1842. Letter to Asa Gray, 26 December, from Louisville, Kentucky. Library, Gray Herbarium, Harvard University, Cambridge, Massachusetts.

SKAGGS, DEBORAH SUSAN. 1982. Charles Wilkins Short: Kentucky botanist and physician. M.A. thesis, University of Louisville. iv, 222 pp.

SMITH, ANNIE MORRILL. 1905. William Starling Sullivant. January 15, 1803 - April 30, 1873. The Bryologist 8: 1-3.

STERNBERG, CHARLES H. 1909. The Life of a Fossil Hunter. H. Holt and Company, New York; Jensen Printing Company, San Diego. xiii, 286 pp.

STOKES, ANSON PHELPS. 1914. William Starling Sullivant (Class of 1823), Born, 1803; Died, 1873, pp. 85-86. *In* Memorials of Eminent Yale Men, Vol. 2: Science and Public Life. Yale University Press, New Haven.

STUCKEY, RONALD L. 1966. The botanical pursuits of John Samples, pioneer Ohio plant collector (1836-1840). Ohio J. Sci. 66: 1-41.

____. 1967. Daniel Steinhauer, early Ohio plant collector, and his correspondence with the botanist Schweinitz. Bartonia 36: 1-24.

____. 1978a. Medical botany in the Ohio Valley (1800-1850). Trans. & Stud. College Physicians Philadelphia 45: 262-279.

____, ed. 1978b. Scientific Publications of Charles Wilkins Short. Arno Press, New York. xi pp. + 22 articles with various paginations.

____. 1979. William Starling Sullivant's contributions to phanerogamic botany in Ohio in the 1830's and 1840's [abstract]. Bot. Soc. Amer. Misc. Ser. Publ. 157: 29-30.

____. 1983. Absence of certain aquatic vascular plants from the Prairie Peninsula, pp. 97-104. *In* Richard Brewer et al., eds. Proceedings of the Eighth North American Prairie Conference. Department of Biology, Western Michigan University, Kalamazoo. viii, 175 pp.

____. 1987a. William Starling Sullivant and his associates [abstract]. Amer. J. Bot. 74: 674.

____. 1987b. First Botanists of Columbus: A Sullivant Family Scrapbook. Compiled . . . for the Field Trip to Green Lawn Cemetery, 38th Annual AIBS Meeting, Columbus, Ohio, 9 August 1987. Ronald L. Stuckey Compiled Publication Series No. 2, Zip Services, Columbus. xiv, 86 pp.

____. 1988. William S. Sullivant and his central Ohio botanical associates. Bartonia 52: 59-73.

____, and **GUY L. DENNY**. 1981. Prairie fens and bog fens in Ohio: Floristic similarities, differences, and geographical affinities, pp. 1-33. *In* Robert C. Romans, ed. Geobotany II. Plenum Press, New York. 263 pp.

____, and **MARVIN L. ROBERTS**. 1982. Monocotyledons, pp. 27-47. *In* Tom S. Cooperrider, ed. Endangered and Threatened Plants of Ohio. Ohio Biological Survey, Biol. Notes No. 16. 92 pp.

____, and **W. ALAN WENTZ**. 1974. Vascular-plant type specimens in the Ohio State University Herbarium. Ohio J. Sci. 74: 20-35. [William Starling Sullivant's 1840 collection of vascular plants from Ohio, pp. 23-24.]

SULLIVANT, JANE D. 1928. Lucas Sullivant -- his personality and adventures. Ohio Archaeol. Hist. Quart. 37: 177-187.

SULLIVANT, JOSEPH. 1838. An Alphabetical Catalogue of Shells, Fossils, Minerals, and Zoophytes, in the Cabinet of Joseph Sullivant, Curator of the Philosophical and Historical Society of Ohio. Published by the author, Columbus, Ohio. 38 pp.

____. 1874. [William Starling Sullivant], pp. 143-160 + Family Table; [Eliza Griscom Wheeler], p. 159; [Joseph Sullivant], pp. 174-184 + Family Table. *In* A Genealogy and Family Memorial. [Published by the author], Columbus, Ohio. 372 pp. + notes and corrections.

SULLIVANT, WILLIAM S. 1838. Letter to John Torrey, 26 November. Historical Society of Pennsylvania, Philadelphia.

____. 1838. Letter to John Torrey, 18 December. Torrey Correspondence, Library, New York Botanical Garden. Printed, 1940. Opp. p. 106. In Rodgers (1940).

____. 1839-1847. Letters to John Torrey, 10 January, 25 April, 9 June, 23 September, 12 December 1839; 24 February, 1 September, 17 October 1840; undated, July 1841; 25 July 1847. Torrey Correspondence, Library, New York Botanical Garden.

____. 1840. Catalogue of the Plants, Native or Naturalized, in the Vicinity of Columbus, Ohio. Charles Scott, Columbus. 63 pp.

____. 1840. Letter to John Torrey, 26 April. Gray Herbarium, Harvard University, Cambridge, Massachusetts.

____. 1840. Letter to Charles W. Short, 27 August. Short Papers, Filson Club, Louisville.

____. 1840-1868. Letters to Asa Gray, 6 January, 27 May, 31 July, 9, 25 November 1840; 8 January, 6 February, 20 April, 18 August, 9 November 1841; undated, early 1842; 1 July, 24 August, 9 December 1842; 19 April, 12 October, 7 December 1843; 19 August, 8 November 1845; 12 July, 12 September, 15 October 1846; 3 November 1850; 11 January, 22 April 1868. Gray Herbarium, Harvard University, Cambridge, Massachusetts.

____. 1842. Account of three undescribed plants of central Ohio. Amer. J. Sci. Arts 42: 49-51.

____. 1848. *Fedia patellaria*, p. 183. *In* Asa Gray. A Manual of Botany of the Northern United States . . . Arranged According to the Natural System. James Munroe & Co., Boston and Cambridge. lxxii, 710 pp.

____. 1849. Letter to George Engelmann, 18 September. Engelmann Papers, Missouri Botanical Garden, St. Louis.

____. 1851. Letter to Asa Gray, 12 October, from Columbus. Library, Gray Herbarium, Harvard University, Cambridge, Massachusetts. Published, 1940. Pg. 208. In Andrew Denny Rodgers III (1940).

____. 1856. Mosses and liverworts, pp. 607-702, 731-737, tab. I-VIII. *In* Asa Gray. Manual of the Botany of the Northern United States . . . Revised Edition. Ivison & Phinney, New York. xxviii, 739 pp. Also published as a separate.

____. 1860. Excerpt from a letter to Edward Tatnall, 19 November. Printed, 1899. Pg. 226. *In* Harshberger (1899).

____. 1866. Letter to Harry Patterson, 21 March. Printed, 1953. Pg. 450. *In* Kibbe, ed. (1953).

____. [undated]. Index to Dewey's "Caricography," Vols. 7-12, 14, 25, 28. 2 pp. manuscript. Mounted in Sullivant's personal copy of John Torrey's "Monograph of North American Cyperaceae" (1836), in the library of The Ohio State University, Columbus.

____, and **LEO LESQUEREUX.** 1864. Icones Muscorum, or Figures and Descriptions of Most of Those Mosses Peculiar to Eastern North America. Sever and Francis, Cambridge, Massachusetts; Trubner & Co., London. [x], 216 pp. + 128 pl. Supplementary ed., 1874. Edited by Leo Lesquereux. With biographical sketch by Asa Gray, pp. 1-8. Charles W. Sever, Cambridge, Massachusetts. Facsimile ed., 1969. With foreword by Andrew Denny Fodgers III, pp. [i-iv]. S-H Service Agency, New York.

SVENSON, H. K. 1932. Monographic studies in the genus Eleocharis--II. Rhodora 34: 193-203, 215-227.

THORNE, C. E. 1909. Joseph Sullivant, p. 617. *In* L. H. Bailey, ed. Cyclopedia of American Agriculture: A Popular Survey of Agricultural Conditions, Practices and Ideals in the United States and Canada, Vol. 4: Farm and Community. Macmillan Co., New York and London. xiv, 650 pp.

TORREY, JOHN. 1836. Monograph of North American Cyperaceae. Ann. Lyc. Nat. Hist. New York 3: 239-448.

____, and **ASA GRAY.** 1838-1843. A Flora of North America: Containing Abridged Descriptions of All the Known Indigenous and Naturalized Plants Growing North of Mexico; Arranged According to the Natural System. 2 vols. Wiley & Putnam, New York and London; Bossange & Co., Paris. xviii, 711 pp.; 504 pp. + index. (Facsimile ed., 1969. Hafner Publishing Co., New York and London.)

TYRRELL, LUCY E. 1987a. A floristic survey of buttonbush swamps in Gahanna Woods State Nature Preserve, Franklin County, Ohio. Michigan Bot. 26: 29-38.

____. 1987b. The rediscovery of *Carex decomposita* in Ohio. Bartonia 53: 15-16.

VOSS, EDWARD G. 1978. [John M. Bigelow (1804-1878)], pp. 56-58. *In* Botanical Beachcombers and Explorers: Pioneers of the 19th Century in the Upper Great Lakes. Contr. Univ. Michigan Herb. 13: i-viii, 1-100.

WALLER, A. E. 1942. Dr. John Milton Bigelow, 1804-1878: An early Ohio physician-botanist. Ohio Archaeol. & Hist. Quart. 51: 313-331. [Concerning Asa Horr, pp. 320-322.]

WARE, JANE. 1991. The botanical angle: OSU professor Ronald Stuckey's interest in early botanist William Starling Sullivant has made Stuckey an expert on Columbus's founding family. Columbus Monthly 16(2): 49-53. February.

WAREHAM, RICHARD T. 1940. New species of mosses and liverworts described by William S. Sullivant and those described by William S. Sullivant and Leo Lesquereux, pp. 302-314. In Rodgers (1940).

WHEATON, J. M. 1882. [W. S. Sullivant enjoys "Carolina parroquet" visiting his home]. P. 405. *In* Report on the birds of Ohio. Ohio Geol. Surv. 4(1): 189-628.

WILSON, JAMES GRANT, and **JOHN FISKE,** eds. 1888. Asa Gray, p. 265; William Starling Sullivant, pp. 742-743. *In* Appleton's Cyclopaedia of American Biography, Vol. 5. D. Appleton and Co., New York. 752 pp.

[**YOUMANS, WILLIAM JAY**]. 1896. Sketch of William Starling Sullivant. Appleton's Popular Sci. Monthly 48: 690-696. (Reprinted, 1896. Pp. 394-401. *In* William Jay Youmans, ed. Pioneers of Science in America: Sketches of Their Lives and Scientific Work. D. Appleton and Co., New York. viii, 508 pp. Facsimile ed., 1978. Arno Press, New York.)

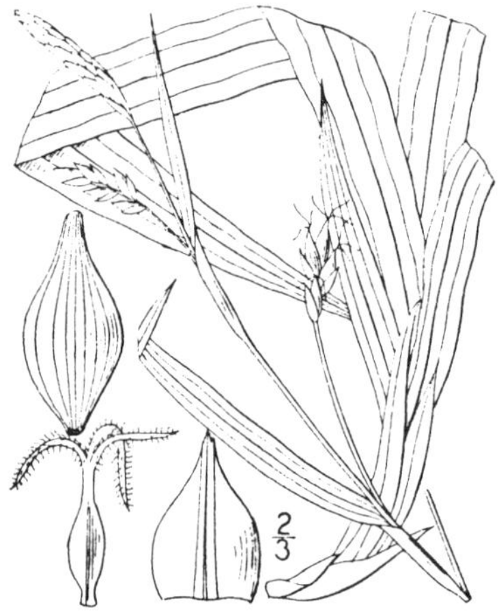

CAREX CAREYANA

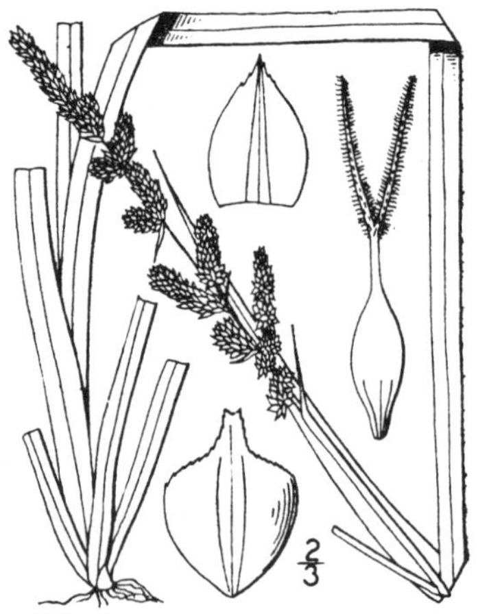

CAREX DECOMPOSITA

APPENDIX I. SULLIVANT'S ASSOCIATES

A. FAMILY ASSOCIATES

JOSEPH S. SULLIVANT (1809-1882), developed an interest in natural history early in life, and prepared during the 1830s, collections of shells, fossils, minerals, zoophytes, and plants of Franklin County. A catalogue of these collections was published (J. Sullivant, 1838). Because of these efforts he had a direct influence of these subjects on his oldest brother, William. Although Joseph's interest in science was ultimately less specialized than his brother's, he was nevertheless an inexhaustible source of encouragement to young and enthusiastic students in any phase of human knowledge provided only that their work was serious and likely to prove worthwhile (Mendenhall, 1915). Joseph S. Sullivant devoted more time to study than to business, as he himself stated. As an organizer of learned societies and educational establishments and a promoter of natural history in central Ohio, he directed his life largely to the promotion of the welfare of the people of Columbus and the state of Ohio. A gentleman of leisure and letters, Joseph had keen interests in art, literature, and science, and was an author on historical, political, scientific, agricultural, and educational subjects.

Joseph S. Sullivant worked diligently to secure a federal land grant college for Franklin County. In 1870, the Agricultural and Mechanical College of Ohio was founded and located on the banks of the Olentangy River north of Columbus. As a charter member and secretary of the board of trustees until 1878, Joseph formulated the first course of study at the new institution, which that same year became The Ohio State University (Kinnison, 1970, pp. 73, 85). He summarized his contributions to society and natural history in an autobiographical account in *A Genealogy and Family Memorial* (1874), the sketch later abridged by Thorne (1909).

ELIZA GRISCOM WHEELER SULLIVANT (1817-1850), the second wife of William Starling Sullivant, was his co-laborer in the writing of descriptions and preparation of illustrations of mosses and liverworts. During the 1840s, she contributed to his success in becoming the foremost nineteenth century authority in North American bryology (Rodgers, 1936, 1940). The second Mrs. Sullivant, a native of New York City, was 14 years younger than her husband. At the time of their marriage on 29 November 1834, the seventeen-year-old bride was somewhat timid and awed by her surroundings at "Sullivant Hill." Her garden was a model to other garden lovers of the community. Many of the live plants had been sent by Asa Gray from the Cambridge Botanic Garden. However, with her simplicity and personal charm, she soon won the love of the people she met in central Ohio.

Despite Mrs. Sullivant's unsteady health, William and Eliza worked and traveled together for 16 years, carrying out their botanical pursuits. They conducted field work in Ohio and in the southern and eastern United States, prepared herbarium specimens of both flowering and non-flowering plants, wrote descriptions in Latin, and prepared exquisite line drawings of mosses and liverworts, some of which were new to science. Many of these illustrations are believed to have been drawn and made into plates by Mrs. Sullivant. These drawings were published in William's books, but according to the practice of the time, they were not credited to him or her. For example, the illustrations of mosses and liverworts published in Sullivant's *Musci and Hepaticae of the United States*, which appeared in the second edition of Gray's *Manual of Botany* (Sullivant, 1856), may have been prepared by Mrs. Sullivant. Some of her drawings were used for many years in the teaching of botany in Heyl's Seminary, Columbus (Rodgers, 1936, p. 96). Of her botanical achievements, William wrote to Asa Gray one and a half years after Eliza's death (Sullivant, 1851):

> But there is an other subject on which it is fitting that to you of whom she always had the highest admiration I should say a few words. I mean her devotion to botanical studies. In these she was my constant companion &

> coworker for the last fourteen years; and I can say with strictest truth that more than half of whatever I have done in these pursuits is due to her. Such a clear and discriminating judgment, such memory, industry, energy and dispatch as were hers is rarely to be met with. She had acquired of Latin & German & French sufficient to read botanical descriptions with ease. In the mechanical labour of writing she was of vast assistance to me. She wrote with remarkable rapidity and accuracy. Indeed accuracy was one of the striking qualities of her mind. In the most difficult & delicate dissections and Microscopical examinations she was astonishingly successful. Her enthusiasm for our studies really required to be checked. Her knowledge of Musci & Hepaticae was very extensive. No female I am sure ever equalled her, and few of the other sex excelled her.

To preserve Eliza's memory as a wife and botanist, William S. Sullivant ordered from Italy a white marble shaft, surrounded by an exquisitely carved bust of his wife, framed in a flowering wreath of *Sullivantia sullivantii*, the saxifragous plant named by Torrey and Gray. The monument (Fig. 18), erected at her gravesite in Green Lawn Cemetery, Columbus, reads:

Eliza G.
wife of
William S. Sullivant
Daughter of
Eliphalet & Mary M.
Wheeler
Born in the City of New York
March 14, 1817
Died in Ohio August 23, 1850

In the *Genealogy and Family Memorial* (1874), her brother-in-law, Joseph Sullivant, described Eliza's accomplishments as a homemaker and botanist:

> She was a lady of much more than common capacity, gifted with a discriminating and well balanced mind, amiable and good tempered; an affectionate wife and tender mother, she managed the affairs of a large household, and, without neglecting any of its cares or duties, still found time to become an expert botanist, and take part in the many frequently extended and delightful botanical rambles, which, after the lapse of many years, cause the writer to look back with vivid and pleasant recollections to the happy times we had at the "Hill" in the days of "Auld Lang Syne" . . . [She was] a lady of rare accomplishment, and, not least, a zealous and acute bryologist, her husband's efficient associate in all his scientific work until her death . . . ."

B. Physician Associates

Dr. Asa Horr (1817-1896), of Baltimore, made extensive collections of plants in the neighborhoods of Baltimore and Pickerington, in Fairfield County, and near Worthington, in Franklin County. He (Fig. 19A) was much interested in grasses and sedges and may have worked with Sullivant in collecting and identifying these plants. Together, they collected water plants. Their association lasted about ten years, when Horr went to Cleveland to attend the Cleveland Medical College, from which he graduated in 1846 (Anonymous, 1894). Later that year Horr moved to Galena, Illinois, and a year later to Dubuque, Iowa, where he continued to practice medicine and also became well-known as an astronomer and meteorologist (Atkinson, 1878; Wilson and Fiske, 1888; Crosby, 1915; Kelly, 1928; Waller, 1942).

Fig. 18. Monument of Eliza Griscom Wheeler Sullivant (photograph by Ronald L. Stuckey, November 1990).

DR. JOHN MILTON BIGELOW (1804-1878), of Lancaster collected plants in Fairfield County (Atkinson, 1878; Waller, 1942; Stuckey, 1978a), and worked with Horr to identify their collections of grasses and sedges. In the summer of 1839 Bigelow became acquainted with Mr. and Mrs. Sullivant, while they were in his neighborhood "herborizing." He gave much credit to the Sullivants for stimulating his efforts in botany (Bigelow, 1843). *The Florula Lancastriensis* . . . (Bigelow and Horr, 1841), published in the *Transactions* of the Medical Convention of Ohio, contained a list of 429 genera and 850 species occurring within the county. A second version was published by Bigelow (1841), and contained notes on medicinal properties.

Bigelow (1843), who wrote a history of botanical exploration in central Ohio, gave high praise to Sullivant, referring to him as an "ardent, intelligent, and scientific cultivator" of botany. He continued:

> About the year 1837, MR. SULLIVANT and his accomplished LADY, in seating themselves upon their beautiful and highly cultivated farm, two miles west of Columbus, found that they were surrounded with the elements and beauty of flora's kingdom. They both determined to enlist under her banner, in determining the extent of Flora's bounty upon their homestead ... [F]rom becoming the amateur botanists of a few acres, they extended their botanical researches, not only over Ohio, but in a measure, over the whole known world. In the short space of time in which he has been engaged in these pursuits, Mr. Sullivant has, probably, personally collected more plants than any other botanist of Ohio.

Bigelow identified Drs. Horr of Baltimore and Paddock of Worthington, as botanists with whom he was personally acquainted, and who, like himself, were "zealously engaged in exploring the [botanical] riches of their immediate neighborhoods, while arduously engaged in the pursuit of their profession, which precludes . . . the opportunity of making their field of observation as broad as they could otherwise wish."

Bigelow (1849) concluded his botanical work in Ohio with an annotated list of 387 medicinal species of the entire state, the last known publication of this type by an Ohio physician. Bigelow became engaged in two important botanical field explorations in the western United States, namely, the Mexican Boundary Survey of 1850 and the Whipple Expedition for the Pacific Railroad Survey of 1853 (Rodgers, 1942; Jepson, 1962). In 1860 he settled in Detroit, where he continued to practice medicine and to write on medical botany, and where he died in 1878 (Noyes, 1878; Burr, 1930; Rodgers, 1942; Voss, 1978).

DR. JONATHAN ROBERTS PADDOCK (1803-1878), a classical scholar and physician, was associated first with the Literary Department and later with the Medical Department of Worthington College, where he taught chemistry, botany, materia medica, and pharmacy. How Paddock and Sullivant became acquainted is not known, but they may have spent many hours botanizing together in central Ohio. They exchanged plant specimens, as evidenced by the presence of some of Sullivant's plants in Paddock's herbarium, which is at the University of Illinois (Anonymous, 1920; Stuckey, 1978a). Their association as botanists was short-lived, however, as Paddock moved to Maysville, Kentucky, following the unexpected closing of the Medical Department of Worthington College in the fall of 1839. In Maysville, he conducted an active private medical practice for the next 20 years, after which he became an invalid and was forced to confine his work to his office until his death in 1878 (Anonymous, 1878a, b).

Paddock (Fig. 19B) "was an excellent chemist, a splendid botanist, and a friend of Mr. Sullivant, of Columbus . . . ," according to Harvey W. Felter (1902, p. 13), the historian of the Worthington Medical College. He further described Paddock as a very skillful physician and classical scholar of extraordinary attainments, "able to name all the plants and trees of the State [of Ohio] by sight" (Felter, 1902, p. 90). While in central Ohio, Paddock documented the vascular flora in the vicinity of Worthington, through his active field work and careful preservation of plant specimens. Unlike his central Ohio botanical associates, however, Paddock did not publish a catalogue of the local flora.

Fig. 19. Botanical associates of William S. Sullivant. **A.** Asa Horr (1817-1896) (from Crosby, 1915, opp. p. 151). **B.** Jonathan Roberts Paddock (1803-1878) (from Felter, 1902, p. 89; also in Felter, 1903, opp. p. 163). **C.** Ichabod Gibson Jones (1807-1857) (from Felter, 1902, p. 89; also in Felter, 1903, opp. p. 168). **D.** Leo Lesquereux (1806-1889) (from McCabe, 1887, opp. p. 845).

DR. ICHABOD GIBSON JONES (1807-1857), another physician in the Medical Department of Worthington College, taught materia medica during the school's early years. He (Fig. 19C) had moved to Columbus in 1834, about the time that Sullivant began his botanical studies. Jones took an interest in scientific botany and is said to have written "several papers descriptive of indigenous plants and trees, of which the most notable, perhaps, is a description of the grasses of this region; and he prepared an herbarium of the flora of central Ohio, the only complete work of the kind of his time" (Loving, 1912). Unfortunately, no such botanical writings have been found, and the last known location of his "complete and beautiful herbarium (still in the possession of his daughter, Miss Emma Jones [d. 9 October 1939, in Columbus])" of indigenous plants was reported by Loving (1904). With his interest in grasses, one wonders if Jones and Sullivant were field companions in the study of these plants. Both men were founding members of the Columbus Horticultural Society and attended meetings together. At the organizational meeting on 12 May 1845, W. S. Sullivant was elected vice-president; Joseph Sullivant, recording secretary; and Jones, one of seven managers (Anonymous, 1845). In March 1848, Jones became the second president of the Society (Dodds, 1959). Emma Jones (1913) wrote that Sullivant said her father "was the finest man he ever knew."

Paddock, Bigelow, Jones, and Horr constituted Sullivant's circle of young physician-botanists. Paddock, who was the same age as Sullivant, was the senior member at age 36 in 1839, quite possibly the last year that all four physicians and Sullivant were in direct association. Bigelow was a year younger, Jones was four years younger, and Horr was fourteen years younger. It is evident from the excerpts presented here that both Horr and Bigelow received much of their botanical assistance and guidance from Sullivant. Horr reportedly attended the Medical Department of the Worthington College from 1836 until his graduation two years later (Anonymous, 1894). While at this school, his early influence in botany may have come from Paddock. Bigelow graduated from the Medical College of Ohio, Cincinnati, in 1832. This school offered no courses in scientific botany, and therefore it would appear that he developed his botanical interests through interaction with Sullivant. Paddock was the only member of this group to hold an early teaching position in a medical school, and thus was in a position to advance botanical knowledge as a part of his profession. He may well have been the most influential member of the group of physicians during their formative years in botany.

C. LEO LESQUEREUX

In 1848, Sullivant hired **LEO LESQUEREUX (1806-1889)**, a largely self-taught Swiss botanist and geologist who was particularly knowledgable on the subjects of mosses, peat bogs, glaciation, and coal formation. Born in Fleurier, Switzerland, Lesquereux initially followed his father's craft of watchmaker. After the Revolution of 1848 and the subsequent abolishment of the Society of Natural History of Neuchatel, he followed the example of his friend Louis Agassiz and made his way to Boston, where he served a short term as curator in Asa Gray's herbarium. By the end of that year, however, he was hired as a field and laboratory assistant by Sullivant. Until his death Columbus would be home to the gentle expatriate, and Gray himself would remark much later that Columbus "was known in every country of the civilized world, not on account of its city hall, but because it was the home of William Sullivant and Leo Lesquereux" (Mendenhall, 1915).

Lesquereux (Fig. 19D) assisted in the preparation of Sullivant's bryological publications and collections of *exsiccatae*. Following Sullivant's death, he edited the supplement to *Icones Muscorum* (1874) and prepared, with Thomas Potts James, a *Manual of the Mosses of North America* (1884). He is also remembered for his major pioneering contributions to American paleobotany, often made for the numerous pioneering geological surveys of the American middle states and western territories. Lesquereux's life and work have been treated extensively by Lesley (1895), Sternberg (1909), Rodgers (1940, 1944, 1962) and Sarton (1942).

APPENDIX II. PLANTS LISTED IN SULLIVANT'S CATALOGUE (1840) MARKED BY HIM AS NATURALIZED

(Names and authors corrected to current usage)

p. 6 Ranunculus repens L.
p. 10 Capsella bursa-pastoris (L.) Medic.
Camelina sativa (L.) Crantz
p. 12 Hypericum perforatum L.
p. 13 Abutilon avicennae Gaertn. =
A. theophrasti Medic.
Sida spinosa L.
p. 15 Lychnis githago L. =
Agrostemma githago L.
Cerastium viscosum L.
p. 17 Rosa rubiginosa L. =
R. eglanteria L.
p. 20 Trifolium pratense L.
Trifolium repens L.
p. 22 Cannabis sativa L.
Urtica dioica L.
p. 24 Chenopodium album L.
p. 25 Polygonum aviculare L.
Polygonum orientale L.
Rumex crispus L.
Rumex obtusifolius L.
p. 29 Achillea millefolium L.
p. 30 Anthemis cotula L.
Arctium lappa L.
p. 31 Helenium autumnale L.
p. 32 Inula helenium L.
Leontodon taraxacum L. =
Taraxacum officinale Weber
p. 33 Sonchus spinulosus Bigel. =
S. asper (L.) Hill
p. 34 Dipsacus sylvestris Huds.
Plantago lanceolata L.
p. 35 Cynoglossum officinale L.
Leonorus cardiaca L.
Marrubium vulgare L.
p. 36 Melissa clinipodium Benth. =
M. officinalis L.
Mentha piperita L.
Nepeta cataria L.
p. 39 Verbascum thapsus L.
Veronica arvensis L.
Veronica officinalis L.
Datura stramonium L.
Solanum dulcamara L.
p. 47 Phleum pratense L.
Pennisetum glaucum Nutt. =
Setaria glauca (L.) Beauv.
Pennisetum viride Nutt. =
Setaria viridis (L.) Beauv.
Agrostis polymorpha Huds. =
A. stolonifera L.
p. 49 Bromus secalinus L.
Dactylis glomerata L.
Poa pratensis L.
p. 50 Triticum repens L. =
Agropyron repens (L.) Beauv.

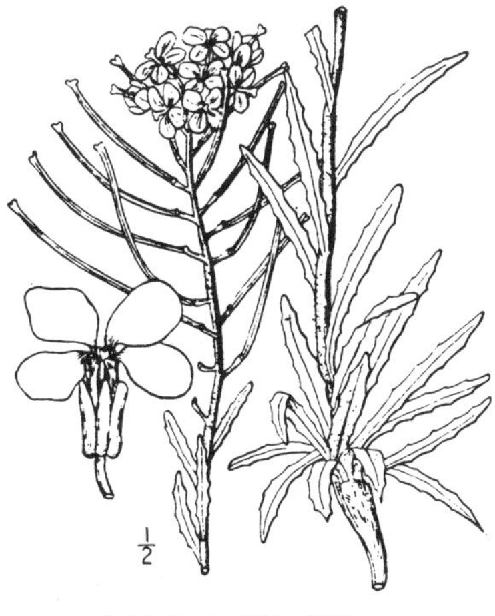

ERYSIMUM ARKANSANUM

APPENDIX III. LIST OF 15 TAXA IN TORREY AND GRAY'S TWO VOLUMES OF THE FLORA OF NORTH AMERICA (1838-1843) BASED ON SPECIMENS COLLECTED BY WILLIAM S. SULLIVANT AND SENT TO TORREY AND GRAY.

(Taxa named and described by them are marked with asterisks.)

  Aster carneus Nees, 2: 133
* Aster laxiflorus Nees τ ? laetiflorus Torrey & Gray, 2: 138
  Aster shortii Hooker, 2: 118
  Aster simplex Willdenow, 2: 131
* Coreopsis discoidea Torrey & Gray, 2: 339
  Erysimum arkansanum Nuttall, 1: 667
* Helianthus cinereus Torrey & Gray ß sullivantii Torrey & Gray, 2: 324
  Helianthus grosse-serratus Martens τ, 2: 326
* Lonicera flava Sims ß Torrey & Gray, 2: 6
* Lonicera parviflora Lamarck ß Torrey & Gray, 2: 7
  Rudbeckia speciosa Wenderworth, 2: 308
* Saxifraga ? sullivantii Torrey & Gray, 1: 575
  Sida dioica Cavanilles, 1: 681
  Silphium pinnatifidum Elliott, 2: 276
* Valeriana ciliata Torrey & Gray, 2: 49

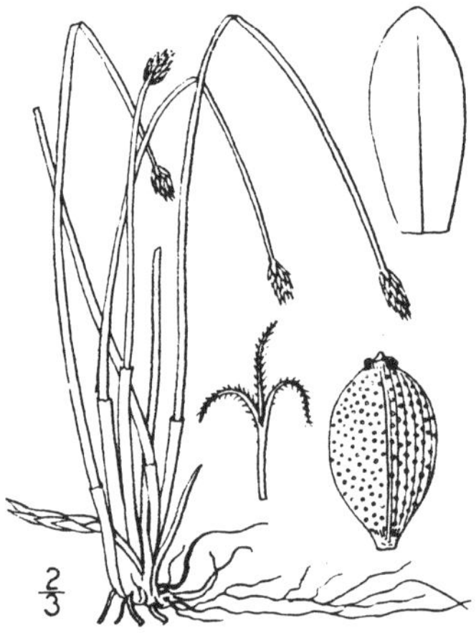

ELEOCHARIS COMPRESSA

ARABIS PATENS

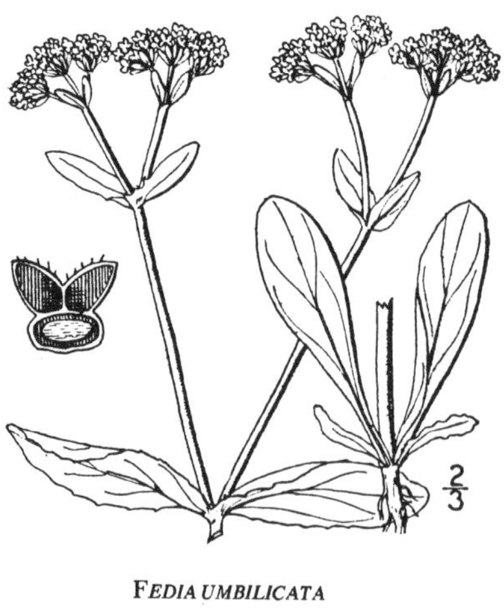

FEDIA UMBILICATA

APPENDIX IV. FLOWERING PLANTS NAMED AND DESCRIBED FROM WILLIAM S. SULLIVANT'S SPECIMENS

SPECIES NAMED BY WILLIAM S. SULLIVANT

ARABIS PATENS Sullivant, Amer. J. Sci. Arts 42: 49. 1842. Type locality: "Rocky banks of the Scioto River, near Columbus, Ohio." Specimens seen: GH, NY, OS, PH.

SPREADING ROCK-CRESS. A locally occurring white-flowered mustard of calcareous rocky habitats along shaded creeks and rivers, ranging from eastern Pennsylvania south to Tennessee and westward to Ohio and Indiana (Hopkins, 1937). Sullivant originally thought his specimens belonged to *Arabis hirsuta*, according to statements in letters written 6 February 1841 to Gray and July 1841 to Torrey. The name "*Arabis hirsuta* Scop. var. *ovata*?" appears on some of Sullivant's herbarium labels. *Arabis patens* has been located in Ohio only from Delaware and Franklin counties. During a study in the spring of 1979 by Anne Filbert of the Division of Natural Areas and Preserves, plants were observed at five different limestone outcrops along the Scioto River from Hayden Run in Franklin County north to just below the O'Shaughnessy Dam in Delaware County. These locations are in the area where Sullivant obtained the original specimens.

ELEOCHARIS COMPRESSA Sullivant, Amer. J. Sci. Arts 42: 50. 1842. Type locality: "Wet places in the Darby Plains, fifteen miles west of Columbus, Ohio." Specimens seen: GH, NY, OS, PH. (*Eleocharis tortilis* Schultz, Sullivant Cat. Pl. Columbus 51, 62. 1840.)

FLAT-STEMMED SPIKE-RUSH. A sedge ranging through the central portion of North America chiefly on calcareous substrates (Svenson, 1932). Sullivant originally treated this taxon as *Eleocharis tortilis* Schultz, as suggested by Torrey and Gray, but he questioned this determination in letters of 12 December 1839 and 17 October 1840 to Torrey and 31 July 1840 to Gray. This name appears in Sullivant's *Catalogue* (1840) and on some of Sullivant's herbarium labels. In Ohio, *Eleocharis compressa* occurs primarily in the calcareous region of the western part of the state on thin wet soil over limestone or dolomite bedrock in abandoned quarries, rock outcrops, wet prairies, and fens. The exact locality where Sullivant obtained his plants in not known, although it evidently was in the wet prairies of Madison County, where the species has not been encountered since. Recent occurrences are known from Adams, Erie, Lucas, Portage, Ottawa, and Summit counties (Stuckey and Roberts, 1982).

FEDIA PATELLARIA Sullivant in A. Gray, Manual, ed. 1, p. 183. 1848. = *Valerianella umbilicata* (Sullivant) Alphon. Wood f. *patellaria* (Sullivant in A. Gray) D. M. Eggers. Type locality: "Low grounds, Columbus, Ohio, *Sullivant*." Specimens seen: GH, NY, PH.

Sullivant recognized this *Fedia* as a new species on the label of the specimen he sent to C. W. Short (PH), where he wrote "*Fedia*-------sp. nov.? I can find nothing in my books that will answer for this -- common here -- Columbus, Ohio W.S.S. 1842."

FEDIA UMBILICATA Sullivant, Amer. J. Sci. Arts 42: 50. 1842. = *Valerianella umbilicata* (Sullivant) Alphon. Wood f. *umbilicata*. Type locality: "Around Columbus, Ohio." Specimens seen: GH, MO, NY, PH.

In Sullivant's day, these two species of *Fedia* were not very common, or perhaps they were quite rare, in the Columbus, Ohio, area. According to Porter (1872), these two species "soon disappeared from their original station, and no botanist seems to have met with either of them . . ." until the spring of 1870, when *F. patellaria* was discovered near Columbia, along the Susquehanna River in Pennsylvania, and the following spring, when *F. umbilicata* was found along the Hudson River in New York. Porter treated the two species as formal varieties of *Fedia radiata* Michx. By the mid-twentieth century these two taxa had spread into the deforested open floodplain valleys of the rivers in central Ohio. In late May and early June, unplowed fields can be seen covered entirely with these plants with their small white flowers. They create a mass of vegetation across the landscape, appearing as if covered by a blanket of greenish-white snow. During the 1960s some of these populations were studied in the field by Donna Marie Eggers-Ware (1970). Ultimately, she reduced Sullivant's two species of *Fedia* to forms of *Valerianella umbilicata*, namely *V. umbilicata* f. *patellaria* and *V. umbilicata* f. *umbilicata* (Eggers-Ware, 1983).

TAXA NAMED BY OTHER BOTANISTS

ASCLEPIAS SULLIVANTII Engelmann in A. Gray, Manual, ed. 1, p. 366. 1848. Type locality: "Near Columbus, Ohio, *Sullivant*. (Also Missouri, *Engelmann*)." Specimens seen: PH (*Sullivant's*), MO (*Engelmann's*).

ASTER LAXIFOLIUS Nees τ ? LAETIFLORUS Torrey and Gray, Fl. N. Amer. 2: 138. 1841. = *A. borealis* Rydberg. Type locality: "Milwaukie, Wisconsin, *Dr. Lapham*! Columbus, Ohio, *Mr. Sullivant*!" Specimens seen: GH, NY.

All of the specimens seen are labeled *Aster laxifolius* by Asa Gray, but in the publication, Torrey and Gray used the varietal epithet *laetiflorus*.

CAREX ALATA Torrey var. FERRUGINEA Fernald, Proc. Amer. Acad. Arts 37: 477, pl. 2. 1902. Type locality: "Ohio to Michigan, Illinois, and Iowa."

CAREX ANGUSTATA Boott var. α TYPICA Boott, Illus. *Carex* 173, pl. 586. 1867. = *C. stricta* Lamarck var. *angustata* (Boott) Bailey. Type locality: "Ohio, *Sullivant*."

CAREX CONJUNCTA Boott, Illus. *Carex* 122, pl. 392. 1862. Type locality: "In America septentrional: . . . Ohio, *Sullivant* . . . The specimens figured are from Columbus, Ohio, *Sullivant*."

CAREX FOENEA Willdenow var. ß Boott, Illus. *Carex* 118, pl. 376. 1962. = *C. suberecta* (Olney) Britton. Type locality: "Ohio, *Sullivant, Lesquereux*."

CAREX LAXIFLORA Lamarck var. ϵ BLANDA Boott, α MAJOR Boott, Illus. *Carex* 37, pl. 92, fig. 1. 1858. Type locality: ". . . Ohio, *Sullivant* . . . ."

CAREX LAXIFLORA Lamarck var. ϵ BLANDA Boott, τ GRACILLIMA Boott, Illus. *Carex* 38, pl. 91, fig. 2. 1858. Type locality: "Ohio, *Sullivant*."

CAREX LAXIFLORA Lamarck var. ζ LATIFOLIA Boott, Illus. *Carex* 38, pl. 93. 1858. d= *C. albursina* Sheldon Type locality: ". . . Ohio, *Sullivant* . . . ."

CAREX SULLIVANTII Boott in A. Gray, Amer. J. Sci. Arts 42: 29. 1842. = *C.* X *sullivantii* Boott in A. Gray. (*C. hirtifolia* Mackenzie X *C. gracillima* Schweinitz = "*C. pubescens* Muhlenberg X *C. gracillima* Schweinitz."). Type locality: "in sylvaticis prope Columbum, Ohionis, ubi detexit . *W. S. Sullivant*, cum *C. pubescente, C. gracillima*, etc. vigens." Specimens seen: GH, MO, NY, PH.

SULLIVANT'S SEDGE. A very rare sedge of hybrid origin, reported from three localities in the northeastern United States (Fernald, 1950; Howe, 1881; Mackenzie, 1935). In Ohio, it has not been known since Sullivant's time, until two specimens were identified in the 1980s from Huron and Muskingum counties. (Huron Co.: Fitchwille Twp., 10 June 1980, *D. M. Brandenburg 486*, KE, OS; Muskingum Co.: Below Taylor's Falls, Sec. 8, Highland Twp., 6 June 1968, *A. W. Cusick 7205*, KE, MU).

COREOPSIS DISCOIDEA Torrey and Gray, Fl. N. Amer. 2: 339. 1842. = *Bidens discoidea* (Torrey and Gray) Britton. Type locality: "Wet places and swamps, Columbus, Ohio, Mr. Sullivant! . . ." Specimens seen: NY, PH. (*Bidens* aff. *frondosa* ?, Sullivant, Cat. Pl. Columbus 30, 47. 1840. "Resembles *B. frondosa* . . . Grows plentifully in the 'Cat-tail' swamp, 3 miles west of Columbus.")

HELIANTHUS CINEREUS Torrey and Gray ß ? SULLIVANTII Torrey and Gray, Fl. N. Amer. 2: 324. 1842. = *H.* X *cinereus* Torrey and Gray in Jackson and Guard (*H. mollis* Lamarck X *H. occidentalis* Riddell subsp. *occidentalis*.). Type locality: "Near Columbus, Ohio, Mr. Sullivant!" Specimens seen: GH, NY.

On one of the specimens that Sullivant sent to Torrey and Gray (GH), he noted that it came from the "Darby Plains 1840 Oct 1st," and in an undated letter to Gray he wrote that "*H. cinereus* was a plant that struck me as something not common, it must be rare -- I never met with (it) but in one spot in the Darby Plains."

JUNCUS BRACHYCARPUS Engelmann, Trans. St. Louis Acad. Sci. 2: 467. 1866. Type locality: "In the Mississippi Valley from central Ohio, *Sullivant*..." Specimen seen: MO.

JUNCUS CANADENSIS J. Gay var. ß BRACHYCEPHALUS Engelmann, Trans. St. Louis Acad. Sci. 2: 473-475. 1866. = *J. brachycephalus* (Engelmann) Buchenau. Type locality: ". . . Central Ohio, *Sullivant* . . . ." Specimen seen: MO.

LONICERA FLAVA Sims ß Torrey and Gray, Fl. N. Amer. 2: 6. 1841. = *L. prolifera* (Kirchner) Rehder. Type locality: "Banks of the Scioto [River] above Columbus, Ohio, *Sullivant*! Milwaukee, Wisconsin, *Mr. Lapham*!" Specimens seen: GH, NY. Honeysuckle: (See note under next species.)

LONICERA PARVIFLORA Lamarck ß ? Torrey et Gray, Fl. N. Amer. 2: 7. 1841. = *L. parviflora* Lamarck ß ? *sullivantii* Wood = *L. sullivantii* (Wood) A. Gray = *L. prolifera* (Kirchner) Rehder. Type locality: "Near Fort Gratiot, Michigan, *Dr. Pitcher*! On limestone cliffs along the Scioto &c. Ohio. *Mr. Sullivant*!" Specimen seen: NY. ("*Lonicera*-------(sp. nov.) ?" Sullivant, Cat. Pl. Columbus 29, 57. 1840.) "Grows with *L. flava*, on the rocky banks of the Scioto, above Columbus."

GRAPE HONEYSUCKLE; SULLIVANT'S HONEYSUCKLE. A twining or sprawling shrub with attractive yellow flowers, this species of dry rocky banks and open woodlands in the interior of the United States was originally described by Torrey and Gray as unnamed variants of *L. flava* Sims and *L. parviflora* Lamarck, based on specimens from Sullivant and others. In his *Catalogue* (1840), Sullivant recognized it as a new species, but did not provide a name. Wood later provided the epithet *sullivantii* as a variant of *L. parviflora*, and Gray raised this taxon to specific rank, creating the name *L. sullivantii*. Today, this name is considered a synonym of *L. prolifera*. Gray (1883) and Sargent (1890) wrote on the early history of the naming of Sullivant's honeysuckle and recorded that Sullivant sent plants to Gray, who then propagated them in the botanic garden at Harvard College, whence they were widely distributed and cultivated in gardens in New England and Europe. In Ohio, the species is known from 15 counties, mostly in the western portion of the state (Braun, 1961; Roberts and Cooperrider, 1982).

RUDBECKIA SULLIVANTII C. L. Boynton and Beadle, Biltmore Bot. Studies 1: 15. 1901. = *R. speciosa* Wenderoth var. *sullivantii* (C. L. Boynton and Beadle) Robinson = *R. fulgida* Aiton var. *sullivantii* (C. L. Boynton and Beadle) Cronquist. Type locality: ". . . Columbus, Ohio, by *Mr. W. S. Sullivant*." Specimens seen: GH, NY, OS. (*Rudbeckia fulgida* Aiton. Sullivant, Cat. Pl. Columbus 32. 1840.)

SAXIFRAGA ? SULLIVANTII Torrey and Gray, Fl. N. Amer. 1: 575-576. 1840. = *Sullivantia ohionis* Torrey and Gray = *Sullivantia sullivantii* (Torrey and Gray) Britton. Type locality: "Limestone cliffs, Highland County, Ohio, *Mr. Wm. S. Sullivant*!"; ". . . limestone cliffs of the rocky fork of Paint Creek, 3/4th of a mile above the junction -- very abundant," as noted by Sullivant on a label of a specimen sent to Short in 1842 (PH). Specimens seen: GH, MO, NY, OS, PH. ("*Heuchera* ? ------- ?" Sullivant, Cat. Pl. Columbus 20, 57. 1840.)

SULLIVANTIA. *Sullivantia sullivantii* occurs on cool, moist, steep, shaded cliffs of either sandstone or calcareous rock; endemic to a region encompassing portions of southern Ohio, Carter County in northeastern Kentucky, and three counties in southeastern Indiana. Sullivant's involvement in the discovery and naming of this species is amply documented in the main text of this paper and in Rodgers (1940, pp. 130-134). A comprehensive report on the biology, habitat, and distribution of the species concluded that it was not endangered or threatened, because many of the species' localities are protected in state parks or state nature preserves (Medley, 1980).

VALERIANA CILIATA Torrey and Gray, Fl. N. Amer. 2: 49. 1841. = *V. edulis* Nuttall in Torrey and Gray subsp. *ciliata* (Torrey and Gray) F. G. Meyer. Type locality: "Swamps and wet alluvial prairies, 'near Springfield, Ohio, *Mr. Williams*,' *Mr. Sullivant*! Urbana, Ohio, *Mr. Samples*! Milwaukie County, Wisconsin, *Mr. Lapham*! . . ." Specimens seen: GH, NY, OS, PH. ("*Valeriana* ------- (sp. nov.)." Sullivant, Cat. Pl. Columbus 34, 59. 1840. "A new and unpublished species of *Valeriana*; found, as I am informed, by Mr. Samples, lately of Urbana, in the Cedar swamp, 6 miles south of that place").

On the label of a specimen sent to Torrey (NY), Sullivant wrote a brief history of this new species: "No 41 *Valeriana* ------- ? I gathered this plant this spring -- could find no description of it in any of my books -- thought it must be new & was in hopes I should be the first to send it to you, but find I have been anticipated. -- My friend T. G. Lea Esq. of Cin[cin]n[a]ti wrote me a few days since, that he had sent you a *Valeriana* from the same neighborhood collected by Mr. Samples w[h]ich was new to you. -- I have since ascertained that it was collected a year or two since by Mr. [Milo G.] Williams of Springfield & sent by him to Dr. Short -- from whom nothing had been heard on the subject. -- . . . [description omitted here] . . . -- Cedar Swamp near Springfield Ohio -- 1839. Wm S. S[ullivant]." Additional details on the history of the discovery of *V. ciliata* are contained in an extensive study of the botanical pursuits of John Samples by Stuckey (1966, pp. 6-9, 37-38). *Valeriana ciliata* continues to be a rare species in Ohio, where recent occurrences are known from Cedar (Swamp) Bog and the Urbana Raised Bog in Champaign County, and Mickey Fen and Liberty Fen west of West Liberty in Logan County (Stuckey and Denny, 1981).

APPENDIX V. SELECTED RARE SPECIES IN OHIO DISCOVERED BY WILLIAM S. SULLIVANT

ARONIA ALNIFOLIA = *Amelanchier sanguinea* (Pursh) DC. Specimen seen: NY.

ROUND-LEAVED SERVICEBERRY. "I have found but one small shrub," as noted by Sullivant in a letter to Torrey, 9 June 1839. Sullivant also noted on an herbarium specimen label of a plant sent to Torrey (NY) that he had ". . . met with but one shrub 2 to 3 feet high on the limestone cliffs on the Scioto [River] Columbus, Ohio." Rare in Ohio where known from five south-central counties and from one locality on Lake Erie as mapped by Braun (1961).

CAREX CAREYANA Torrey ex Dewey. Specimens seen: GH, MO, NY, OS, PH.

CAREY'S SEDGE. "Grows abundantly on the hill side, three miles west of Columbus, immediately north of the National Road. I have heard of but one other locality for this plant. T. G. Lea, Esq., of Cincinnati, informs me that 2 or 3 years since, he found a few specimens of it, in his neighborhood, but that subsequent examinations for it, have proved unavailing." It is "one of the earliest in flower--does not 'grow in marshes' here--its locality here is a dry woody hill side" (Sullivant, Cat. Pl. Columbus, Note 40, p. 62. 1840). Local in Ohio where mapped for 15 counties by Braun (1967).

CAREX DECOMPOSITA Muhl. Specimens seen: GH, MO, OS, PH.

CYPRESS-KNEE SEDGE. This species is of local occurrence in the eastern and midwestern United States, and, as mapped by Stuckey (1983), is one of several species that shows a pattern of distribution absent from the Prairie Peninsula. Sullivant provided the first known locality in Ohio for this species, as noted on the label of the specimen he sent to Engelmann (MO):

> ". . . Rare plant grows in the cattail swamp 3 miles west of Columbus -- Ohio Collected 1838 & 39."

According to Braun (1967), *Carex decomposita* is known in Ohio "only from Cranberry Island in Buckeye Lake," Licking County, but this rare species should also be credited to Franklin County from Sullivant's specimens of 1838 and 1839 and to Hamilton County from Lea's specimens of 1842 (Braun, 1934). Specimens were last obtained in Ohio at Cranberry Island in 1910 (Stuckey and Roberts, 1982), until the species was rediscovered in November 1984 during a floristic survey of buttonbush swamps in the Gahanna Woods State Nature Preserve, Franklin County (Tyrrell, 1987a, b).

ERYSIMUM ARKANSANUM Nutt. = *Erysimum capitatum* (Dougl.) Greene. Specimens seen: GH, NY, OS, PH.

WESTERN WALLFLOWER. ". . . First detected in this county in the summer of 1838 by my brother J[oseph] Sullivant" (Sullivant, Cat. Pl. Columbus, Note 2, p. 57. 1840). Not known in Ohio since Sullivant's record of 1838, until 1977 when this western tall yellow- to orange-flowered mustard of dry prairies, sandy soils, and open woodlands was rediscovered east of Aberdeen in Brown County by Allison W. Cusick (19, 352, KE, OS). Here at the eastern-most edge of its range, *Erysimum arkansanum* may represent a relic of the dry prairie flora that flourished more extensively in Ohio during the Xerothermic Period, eight to five thousand years ago. The species has been treated under the name of *E. capitatum*

(Dougl.) Greene by Rossbach (1958, 1963), with which it is synonymous, and incorrectly under the name *E. asperum* (Nutt.) DC., a species of the Great Plains, by Easterly (1964).

EULOPHUS AMERICANUS Nutt. = *Perideridia americana* (Nutt. ex DC.) Reichenb. Specimens seen: GH, NY, PH.

Sullivant's knowledge of this species is well preserved on labels of specimens he sent to Torrey (NY) and to Gray (GH).

> "No. 91 This is new to me. -- I found an imperfect specimen of it last year. -- grows upon dry ground in the 'Barrens' of 'Darby Plains' a few miles north of this [?] -- scarce -- unable to procure riper fruit -- nothing peculiar in the root. -- Wm S. S[ullivant] -- Columbus O[hio] Sept 1840."

In a letter to Torrey, 17 October 1840, Sullivant stated:

> "An other umbelliferous plant about which I feel some interest, is No 91, since I sent that I have seen the same plant (in flower only) in Dr Shorts herbarium he found but a single imperfect specimen of it 3 or 4 years ago -- never has been able to meet with it since -- was unable to name it --"

On a specimen label in Gray's herbarium (GH), Sullivant wrote at length of the history of his sending plants of this species to Torrey:

> "Of this plant Dr Torrey in litt. Nov. 13th 1840 -- 'The umbelliferous plant (No. 91) is not described in the flora -- It is new to me & I should be greatly pleased to receive specimens of it with ripe fruit.' I sent this umbelliferous plant first one specimen June 18th 1839 -- again in June 1840 four or five specimens -- again June 15th 1841 six or eight specimens -- always in dry meadows 1/2 to 1 mile east of Lafayette Madison Co[unty] Ohio -- appears to be very rare & delicate -- soon dies off -- & by middle of July scarcely a vestage to be seen of it -- Root fasiculately tuberous -- . . . Very little smell or taste of a peculiar kind -- the taste is very marke[d]ly that of a carrot or rather more that of a sweet potatoe, neither has the seed (immature) any peculiar flavor. July 4th 1841 -- Examined the fruit of the new umbelliferous plant, which is now ripe but not consoledated . . . ."

On another specimen label (GH), Sullivant commented in 1841 to Gray:

> ". . . This plant is extremely rare I have been able to collect but very few specimens the half of which I send to you. W. S. S[ullivant]."

In a letter of July 1841 to Torrey (NY), Sullivant further explained his problems with this species:

> "Enclosed is some entirely ripe & perfect fruit of the new umbelliferous plant (Neurophyllum No. 2)! -- which I have procured at last with no little trouble -- the plant is extremely scarce -- it has disappeared from its old locality. The trouble of it has not ended here with me -- its persuit in the books has been more unsuccessful than in the field . . . But the best disposition (as I think) of it is to consider it . . . near if not belonging to Nuttall's *Eulophus* -- from description *Eu[lophus]* *Americanus* must look much like our plant . . . ."

On the label of Sullivant's specimen sent to Short (PH) is the note: "*Eulophus Americanus*, Nutt. Hab. Darby Plains 20 miles west of Columbus, O[hio] '*planta rariforma*'!!! W. S. Sullivant 1842."

Perideridia americana has not been seen in Ohio since Sullivant's time, except by Floyd Bartley who obtained a specimen in 1950, two miles southeast of Whisler, Pickaway County (Chuang and Constance, 1969). Here at the eastern edge of its range, this species may represent a relic of the dry prairie flora that flourished more extensively in Ohio during the Xerothermic Period, eight to five thousand years ago.

NOTHOSCORDUM BIVALVE (L.) Britt. Specimens seen: GH, OS.

FALSE GARLIC. Ranging from Mexico and Texas eastward into the southeastern United States, this species extends northward into southern Ohio where it has been located in sandy meadows and at the edge of open woods. Known in the 1830s from Champaign County and in the 1840s from Sullivant's specimens collected in Clark County, and possibly in the 1820s from Muskingum, Ross, or Tuscarawas counties, but not found in any of these counties since, and in 1932 and 1978 and more recently from Adams County. A complete history of the records of the occurrence of this species in Ohio is in Stuckey (1966, pp. 9-10; 1967, pp. 15-16).

Sullivant's knowledge of this species is recorded on a note with the label of a specimen sent to Gray (GH):

> ". . . *Allium striatum*, Ell? -- I found this 1st. of last May near Springfield Oh[io] -- I have since ascertained that (more than probable) the same was found last season by Mr. Samples near the same place it has none of the taste or smell of the onion -- generally larger than these specimens, 8 to 10 inches. -- Sept 1840 Columbus Ohio Wm S. S[ullivant]."

Nothoscordum bivalve had not been seen in Clark County since Sullivant's time, until discovered on 8 May 1987 by Guy L. Denny (1987), the label for whose specimen reads "Springfield Fen State Nature Preserve, near Springfield, Ohio" (*Guy L. Denny 100*, OS).

EULOPHUS AMERICANUS

NOTHOSCORDUM BIVALVE

APPENDIX VI. LIST OF PLANTS COLLECTED DURING 1840 BY WILLIAM S. SULLIVANT FROM A LETTER BY W. S. SULLIVANT TO CHARLES W. SHORT, 27 AUGUST 1840.

73 taxa, 22 specimens, only *PINUS CANADENSIS* appearing in Sullivant's Catalogue (1840).
* = Specimen in Sullivant collection at OS.
PH = Specimen at the Academy of Natural Sciences of Philadelphia (Short Herbarium).
-- = Plants seen along the sandy shore of Lake Erie or in the tamarack bogs of northeastern Ohio.

Acer dasycarpum = A. saccharinum
* Allium striatum ? = Nothoscordum bivalve
-- Aira flexuosa = Deschampsia flexuosa
Aira triflora
Asplenium ebeneum = A. platyneuron
-- Acerates longifolia = A. floridana
-- Andromeda calyculata = Chamaedaphne calyculata
-- Apocynum pubescens = A. cannabinum
Atheropogon apludoides = Bouteloua curtipendula
-- Botrychium fumarioides
-- * Brasenia peltata = B. schreberi
-- Carex buxbaumii
Carex aurea
Carex aquatilis ?
Carex conoidea
Carex miliacea = C. prasina
-- Cakile maritima = C. edentula
Capraria multifida = Leucospora multifida
Caulina fragilis = Najas flexilis
-- Chrysanthemum leucanthemum
-- * Comarum palustre = Potentilla palustris (PH, Sphagnous swamp northern parts of Ohio, 1842)
-- * Cyperus schweinitzii (PH, Sandy shores of Lake Erie, Cleveland)
-- Eriophorum angustifolium
-- * Euphorbia polygonifolia (PH, Sandy shores of Lake Erie, Cleveland)
* Glyceria acutiflora
Gnaphalium uliginosum
* Geranium robertianum
* Helianthemum canadense
Hydrocotyle americana
Juniperus communis
Juncus bufonus
Kuhnia eupatorioides
Lemna polyrhiza = Spirodela polyrhiza
Linum virginianum
-- Lonicera parviflora var. ? = L. prolifera
Luzula pilosa = L. acuminata
* Lycopodium apodum = Selaginella apoda
-- * Myriophyllum heterophyllum (PH, Akron Summit)
Myriophyllum spicatum = M. exalbescens
* Myriophyllum verticillatum
* Oxycoccus macrocarpus = Vaccinium macrocarpon
Phaca neglecta = Astragalus neglectus
Pinus canadensis = Tsuga canadensis
Poa annua
* Polanisia graveolens = P. dodecandra
Potentilla anserina
Potentilla arguta
Pontederia cordata
Pyrola secunda
Ranunculus acris
Ranunculus pennsylvanicus
Ranunculus sceleratus
Rensselaria virginica = Peltandra virginica
Rubus odoratus
Rubus obovatus = R. hispidus var. obovalis
-- Rubus saxatilis var. canadensis
Rubus strigosus = R. americanus
-- * Sarracenia purpurea
Scirpus planifolius = S. verecundus
-- Shepherdia canadensis
-- * Silene nivea (PH, Zoar)
Spiraea tomentosa
Sporobolus asper
Symphoria racemosa = Symphoriocarpus albus
* Tiarella cordifolia
* Tricuspis seslerioides = Tridens flavus
* Udora canadensis = Elodea canadensis
-- Uralepsis aristulata = Triplasis purpurea
Urtricularia setacea = U. subulata
* Viburnum lentago
-- * Vicia caroliniana (PH, Lancaster)
-- * Viola muhlenbergia = V. labradorica (PH, Cedar Swamp, Urbana)
-- Zannichellia palustris

APPENDIX VII.
WILLIAM S. SULLIVANT'S DOCUMENTED FLOWERING PLANT COLLECTING LOCALITIES IN OHIO AND THE SPECIMENS SEEN FROM THOSE PLACES

The localities are marked on the map of the state of Ohio (Fig. 12).

1. Champaign County
   a. Cedar Swamp near Springfield
      Valeriana ciliata, in 1839 (NY), in 1840 (OS), in 1840, 1842 (GH), 18 May 1842 (PH)
      Viola adunca (PH)
   b. Cedar Swamp, 6 mi. south of Urbana; 50 mi. from Columbus
      Betula pumila
      Rhamnus alnifolia (PH)
      Silphium terebinthinaceum
      Solidago ohioensis
      Solidago riddellii
      Utricularia intermedia (NY), (W. S. Sullivant, Cat. Pl. Columbus, Note 22, p. 60. 1840).
   c. Urbana
      Rubus triflorus = *R. pubescens* (GH), in 1840 (PH)
      Viola muhlenbergii = *V. conspersa*, in 1840 (PH)
2. Clark County, near Springfield
      Allium striatum = *Nothoscordum bivalve*, in 1840 (GH, OS).
      Salix interior, in 1840 (PH)
      Micromeria nuttallii = *Satureja arkansana*, in 1840 (PH)
      Taxus canadensis, in 1840 (PH)
3. Cuyahoga County, sandy shores of Lake Erie, Cleveland
      Cyperus schweinitzii, in 1840 (PH)
      Euphorbia polygonifolia, in 1840 (PH)
4. Erie County, Sandusky
      Satureja arkansana, in 1840 (PH)
5. Fairfield County, Lancaster
      Corydalis sempervirens, in 1840 (PH)
      Magnolia acuminata, in 1840 (PH)
      Vicia caroliniana, in 1840 (PH), cultivated
      Vitis labrusca, in 1840 (PH)
6. Franklin County
   a. Cat-tail swamp three miles west of Columbus
      Carex decomposita (NY), in 1838, 1839 (MO), in 1840 (GH, OS, PH).
      Coreopsis discoidea = *Bidens discoidea*, "in wet places" (W. S. Sullivant, Cat. Pl. Columbus, Note 5, p. 57. 1840); in 1839, 1842 (NY), in 1840, 1842 (PH)
   b. Columbus
      Asclepias sullivantii (MO), in 1840 (PH)
      Aster laxiflorus [var.] *laetiflorus* = *A. borealis*, in 1839 (NY), in 1842 (GH)
      Carex hirtella, "Waste fields;. . .first discovered by Lyne Starling, Jr." (W. S. Sulliv. ms. 1841).
      Carex tetanica (GH), "wet places," in 1839 (NY)
      Fedia patellaria = *Valerianella patellaria* (GH, NY), in 1840, "common here," in 1842 (PH)
      Fedia umbilicata = *Valerianella umbilicata* (MO), in 1841 (NY), in 1842 (GH, PH)
      Rudbeckia sullivantii = *R. fulgida* var. *sullivantii*, in 1840 (GH, NY, OS)
   c. Hillside, three miles west of Columbus, immediately north of the National Road [= Sullivant Hill]

Carex careyana (MO, PH), in 1839 (NY), in 1840 (GH, OS). (W. S. Sullivant, Cat. Pl. Columbus, Note 40, p. 62. 1840.)

d. O'Harra's Saw mill, along the Scioto River four miles north of Columbus

Erysimum arkansanum = *E. captitatum*, in 1839 (NY), in 1840 (GH, OS, PH), "... first detected in this county in the summer of 1838 by my brother J[oseph] Sullivant" (W. S. Sullivant, Cat. Pl. Columbus, Note 2, p. 57. 1840).

e. Rocky limestone banks along Scioto River north of Columbus

Aronia alnifolia = *Amelanchier sanguinea* (NY); "I have found but one small shrub" (Letter to J. Torrey, 9 June 1839).

Arabis patens, in 1840 (GH, NY, OS, PH)

Lonicera flava = *L. prolifera*, in 1839 (NY: *L. sullivantii* = *L. prolifera*), in 1840 (GH); *Lonicera* "intermediate between *Lonicera hirsuta* and *Lonicera flava* . . . Grows with *L. flava* . . ." (W. S. Sullivant, Cat. Pl. Columbus, Note 4, p. 57. 1840).

Trillium nivale, in 1839 (NY)

f. Meadows immediately west of Franklinton

Orchis leucophaea = *Habenaria leucophaea* = *Platanthera leucophaea*; "plentifully" (W. S. Sullivant, Cat. Pl. Columbus, Note 25, p. 60. 1840).

g. Wet prairie, 2-3 mi. south of Columbus (= "Sullivant's Prairie")

Cacalia ovata = *C. tuberosa* = *C. plantaginifolia* (W. S. Sullivant, Cat. Pl. Columbus, Note 7, p. 57. 1840).

Conoclinium coelestinum = *Eupatorium coelestinum*; "Very rare here; collected by L[yne] Starling, Jr. Borders of prairie, 3 miles south of Columbus" (W. S. Sullivant, Cat. Pl. Columbus, Note 9, p. 58. 1840).

Zygadenus glaucus; "Rare with us. Found by L[yne] Starling, Jr." (W. S. Sullivant, Cat. Pl. Columbus, Note 27, p. 61. 1840).

7. Highland County, Rocky Fork Creek, 0.75 mi. above junction of Paint Creek; 80 mi. south of Columbus

Saxifraga ? sullivantii = *Sullivantia sullivantii*, in 1839 (NY), in 1840, "very abundant" in 1842 (PH); ". . . the only place I have met with it--growing in the crevices of very high limestone cliffs over Paint Creek" (Letter to J. Torrey, 6 January 1840).

Taxus canadensis (PH)

8. Madison County

a. Barrens, 20 mi. west of Columbus

Baptisia leucantha; ". . . have never seen this growing except in the 'Barrens'. . ." (Letter to A. Gray, 19 April 1843).

Carex sullivantii = *C.* X *sullivantii* (MO, NY, PH), in 1842 (GH), ". . . in the extensive barrens west of us [Columbus] . . ." (Letter to A. Gray, 20 April 1841). (*C. hirtifolia* X *C. gracillima* = *C. pubescens* X *C. gracillima*)

Zizia cordata (PH) = *Z. aptera*

Other species listed from the "Barrens" (Letters to A. Gray, 1841, 1842):

Aster azeureus = *A. oolentangensis*
Eulophus americanus = *Perideridia americana*
Helianthus cinereus
Helianthus occidentalis
Silphium pinnatifidum = *S. laciniatum* x *S. terebinthinaceum*
Silphium terebinthinaceum

b. Darby Plains, 15-20 mi. west of Columbus

Eleocharis compressa, in 1840 (GH, NY, OS, PH)

Gentiana puberula = *G. puberulenta*, in 1839 (PH)

Helianthus cinereus var. *sullivantii* = *H.* x *cinereus* (NY), in 1840 (GH), "not common, it must be rare--I never met with [it] but in one spot in the Darby Plains" (Letter to A. Gray, undated). (*H. mollis* X *H. occidentalis*)

Liatris sp., in 1840 (PH)
Silphium pinnatifidum = *S. laciniatum* x *S. terebinthinaceum* (Letter to A. Gray, 24 August 1842).
Silphium terebinthinaceum (Letter to A. Gray, 24 August 1842).

c. 0.5-1.0 mi. east of Lafayette
Eulophus americanus = *Perideridia americana*, in 1840 (NY), in 1841 (GH), "planta rariforma!!!" in 1842 (PH)

9. Summit County, Akron Summit
Myriophyllum heterophyllum, in 1840 (PH)
Vaccinium constablaei = *V. corymbosum* L., in 1840 (NY), in 1841 (GH)

10. Tuscarawas County, Zoar
Silene nivea, in 1840 (PH)

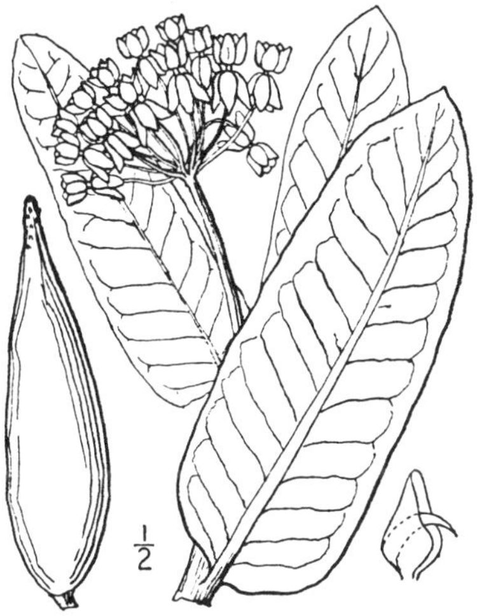

ASCLEPIAS SULLIVANTII

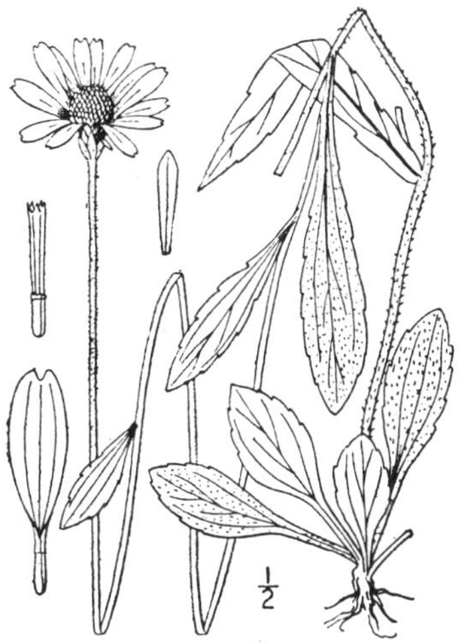

RUDBECKIA SULLIVANTII

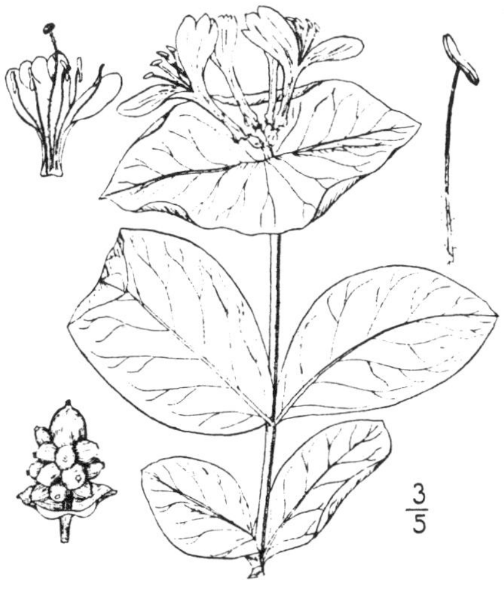

LONICERA SULLIVANTII

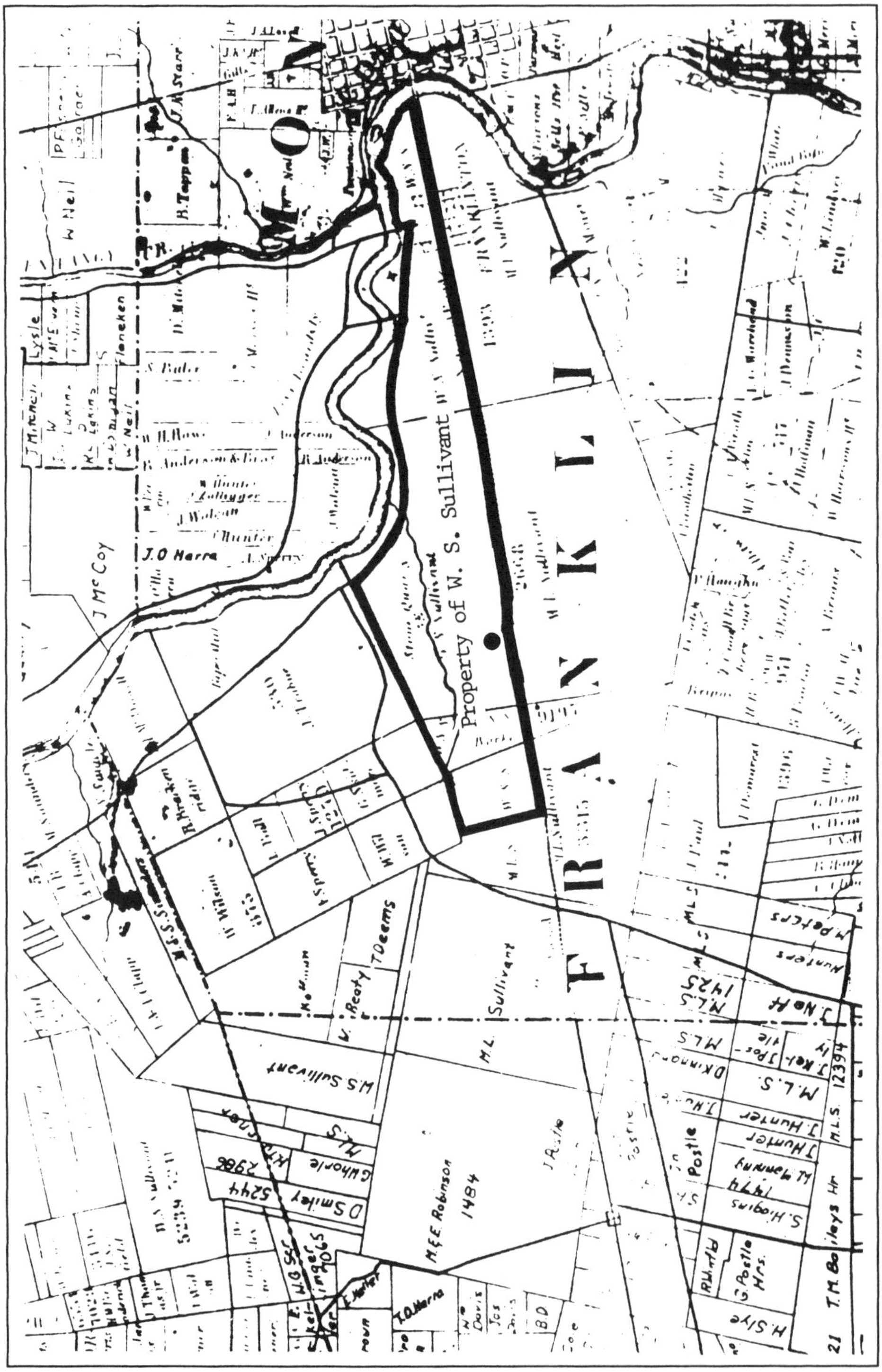

"Sullivant Hill" in Franklin Township located on portion of 1842 map of Franklin County, Ohio.

Index

Combined index with **common plant names** (bold), *scientific names* (italic), and PERSONAL NAMES (small caps).

The text pages of this book were
printed on acid-free recycled paper.